W9-BYA-855

Kelley Wingate

Algebra

Grades 5–9

Credits
Editors: Amy R. Gamble, Elise Craver
Copy Editor: Beatrice Allen

Visit *carsondellosa.com* for correlations to Common Core, state, national, and Canadian provincial standards.

PO Box 35665
Greensboro, NC 27425 USA
carsondellosa.com

© 2014, Carson-Dellosa Publishing, LLC. The purchase of this material entitles the buyer to reproduce worksheets and activities for classroom use only—not for commercial resale. Reproduction of these materials for an entire school or district is prohibited. No part of this book may be reproduced (except as noted above), stored in a retrieval system, or transmitted in any form or by any means (mechanically, electronically, recording, etc.) without the prior written consent of Carson-Dellosa Publishing, LLC.

Printed in the USA • All rights reserved.

ISBN 978-1-4838-0505-4
03-103151151

Table of Contents

Introduction 3

Common Core State Standards Alignment Chart 4

Real Numbers

Adding Real Numbers 5

Subtracting Real Numbers 8

Multiplying Real Numbers 11

Dividing Real Numbers 14

Order of Operations 17

Real-Number Operations with Absolute Value 20

Expressions and Equations

Combining Like Terms 23

Solving One-Step Equations (Addition and Subtraction) 26

Solving One-Step Equations (Multiplication and Division) 29

Solving Basic Multistep Equations 32

Solving Equations with Variables on Both Sides 35

Writing and Solving Equations 38

Solving Inequalities 41

Polynomials

Multiplication and Powers of Exponents 44

Division of Exponents 47

Negative and Zero Exponents 50

Adding and Subtracting Polynomials 53

Multiplying Polynomials and Monomials 56

Multiplying Binomials 59

Factoring

Factoring Polynomials 62

Factoring Polynomials—Special Cases 65

Solving Equations by Factoring 68

Writing and Solving Quadratic Equations 71

Rational Expressions and Ratios and Proportions

Dividing Polynomials 74

Operations with Rational Expressions 77

Ratios and Proportions 80

Graphing and *X*- and *Y*-Intercepts

Graphing Ordered Pairs 83

Graphing Linear Equations 86

Writing Linear Equations 89

Graphing Linear Inequalities 92

Solving Systems of Equations by Graphing 95

Solving Systems of Equations 98

Radicals

Operations with Radicals 101

Answer Key 104

© Carson-Dellosa • CD-104632

Introduction

Competency in math skills creates a foundation for the successful use of math principles in the real world. Practicing algebra skills is the best way to improve at them.

This book was developed to help students practice and master a variety of algebraic concepts. The practice pages can be used first to assess proficiency and later as basic skill practice. The extra practice will help students advance to more challenging math work with confidence. Help students catch up, stay up, and move ahead.

Common Core State Standards (CCSS) Alignment

This book supports standards-based instruction and is aligned to the CCSS. The standards are listed at the top of each page for easy reference. To help you meet instructional, remediation, and individualization goals, consult the Common Core State Standards alignment chart on page 4.

Leveled Activities

Instructional levels in this book vary. Each area of the book offers multilevel math activities so that learning can progress naturally. There are three levels, signified by one, two, or three dots at the bottom of the page:

- Level I: These activities will offer the most support.
- Level II: Some supportive measures are built in.
- Level III: Students will understand the concepts and be able to work independently.

All children learn at their own rate. Use your own judgment for introducing concepts to children when developmentally appropriate.

Hands-On Learning

Review is an important part of learning. It helps to ensure that skills are not only covered but are internalized. The flash cards at the back of this book will offer endless opportunities for review. Use them for basic skill and enrichment activities to reinforce basic algebraic concepts.

There is also a certificate template at the back of this book for use as students excel at daily assignments or when they finish a unit.

Common Core State Standards Alignment Chart

Common Core State Standards*		Practice Page(s)
Operations and Algebraic Thinking		
Write and interpret numerical expressions.	5.OA.1–5.OA.2	17–19
Geometry		
Graph points on the coordinate plane to solve real-world and mathematical problems.	5.G.1	83, 84
Ratios and Proportional Relationships		
Understand ratio concepts and use ratio reasoning to solve problems.	6.RP.3	81, 82
Analyze proportional relationships and use them to solve real-world and mathematical problems.	7.RP.1, 7.RP.3	80–82
The Number System		
Apply and extend previous understandings of numbers to the system of rational numbers.	6.NS.5–6.NS.7	20–22, 83–85
Apply and extend previous understandings of operations with fractions to add, subtract, multiply, and divide rational numbers.	7.NS.1–7.NS.2	5–16
Expressions and Equations		
Apply and extend previous understandings of arithmetic to algebraic expressions.	6.EE.1–6.EE.3	17–19, 23–25, 38–40, 56–58
Reason about and solve one-variable equations and inequalities.	6.EE.6–6.EE.8	41–43, 85
Use properties of operations to generate equivalent expressions.	7.EE.1	53–67
Solve real-life and mathematical problems using numerical and algebraic expressions and equations.	7.EE.3, 7.EE.4	38–40, 71–73
Work with radicals and integer exponents.	8.EE.1	44–52
Understand the connections between proportional relationships, lines, and linear equations.	8.EE.6	86–91
Analyze and solve linear equations and pairs of simultaneous linear equations.	8.EE.7, 8.EE.8	26–40, 95–100
Functions		
Define, evaluate, and compare functions.	8.F.1–8.F.3	85–91
Number and Quantity: The Real Number System		
Extend the properties of exponents to rational exponents.	HSN-RN.2	101–103
Algebra: Seeing Structure in Expressions		
Interpret the structure of expressions.	HSA-SSE.2	59–67
Write expressions in equivalent forms to solve problems.	HSA-SSE.3	68–73
Algebra: Arithmetic with Polynomials and Rational Expressions		
Perform arithmetic operations on polynomials.	HSA-APR.1	53–61
Use polynomial identities to solve problems.	HSA-APR.4	59–61
Rewrite rational expressions.	HSA-APR.6, HSA-APR.7	74–79
Algebra: Reasoning with Equations and Inequalities		
Solve equations and inequalities in one variable.	HSA-REI.3, HSA-REI.4	26–37, 41–43, 68–73
Solve systems of equations.	HSA-REI.6	95–100
Represent and solve equations and inequalities graphically.	HSA-REI.10–HSA-REI.12	85, 92–97

* © Copyright 2010. National Governors Association Center for Best Practices and Council of Chief State School Officers. All rights reserved.

© Carson-Dellosa • CD-104631

Name _____ 7.NS.A.1d

Adding Real Numbers

A number line is a great way to model the addition of real numbers.

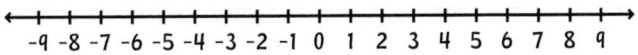

Add 3 + 5. (Start at 3 and move 5 places to the right since 5 is positive.) The answer is 8.
Add 3 + (⁻5). (Start at 3 and move 5 places to the left since 5 is negative.) The answer is ⁻2.
When adding a number that is positive, move to the right. When adding a number that is negative, move to the left.

Explain how to add the following numbers using a number line, and tell the result.

1. 7 + 8 _____

2. ⁻5 + 9 _____

3. 10 + (⁻7) _____

4. ⁻11 + (⁻6) _____

Use the number line to add the numbers.

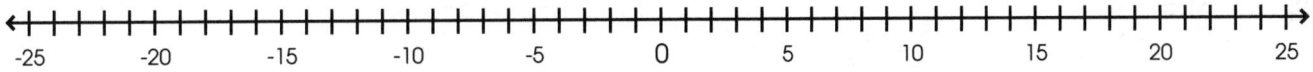

5. 4 + 9

6. ⁻5 + 13 + (⁻11)

7. ⁻7 + (⁻6)

8. 14 + 10 + (⁻3)

9. 0 + (⁻5)

10. ⁻6 + (⁻9) + 10

11. ⁻3 + 3

12. 8 + (⁻2) + (⁻4) + 6

13. 7 + (⁻12)

14. ⁻11 + 5 + (⁻13) + 8

15. ⁻11 + 8 + (⁻1)

16. 12 + (⁻6) + (⁻14) + 17

© Carson-Dellosa • CD-104632

Name _____ 7.NS.A.1d

Adding Real Numbers

$^{-}7 + 6 = {}^{-}1$

Add.

1. $^{-}55 + (^{-}8) + (^{-}4) + 54 =$

2. $1 + (^{-}5) + (^{-}5) + 1 =$

3. $2.7 + (^{-}4.8) =$

4. $3.54 + 4.27 + 7.43 =$

5. $10 + 7 + (^{-}7) + (^{-}10) =$

6. $16 + 21 + (^{-}3) + 7 =$

7. $10 + 7 + (^{-}16) + 9 + (^{-}30) =$

8. $5.8 + 8.4 =$

9. $2.76 + (^{-}6.56) + (^{-}9.72) =$

10. $8 + (^{-}7) =$

11. $2\frac{3}{5} + 4\ 3\frac{3}{7} =$

12. $^{-}8\frac{3}{5} + 3\frac{3}{7} =$

13. $2 + 5 + (^{-}3) =$

14. $^{-}5\frac{3}{4} + (^{-}2\frac{3}{4}) + 8 =$

15. $7.3 + (3.9) =$

16. $^{-}21 + 12 + (^{-}1) + (^{-}17) =$

17. $7.867 + (^{-}5.329) =$

18. $^{-}2\frac{3}{5} + (^{-}5\frac{3}{7}) + 3 =$

19. $3 + 12 + (^{-}13) + 36 =$

20. $^{-}3\frac{1}{6} + (^{-}9\frac{3}{12}) + 6 =$

© Carson-Dellosa • CD-104632

Name _____ 7.NS.A.1d

Adding Real Numbers

Add.

1. $2\frac{3}{5} + (^{-}3\frac{2}{5}) + {}^{-}6 =$

2. $21 + 9 + (^{-}6) + 7 =$

3. $12 + (^{-}9) + 17 =$

4. $2.54 + {}^{-}5.87 + {}^{-}32.65 =$

5. $1.45 + 2.65 + (^{-}9.43) =$

6. $21 + 3 + (^{-}13) + 22 =$

7. $3 + (^{-}3) + 4 + (^{-}5) =$

8. $3.3 + (^{-}3.4) + 5.5 =$

9. $3.6 + (^{-}2.5) + {}^{-}5.5 =$

10. ${}^{-}0.6 + (^{-}0.56) + 3 =$

11. $3\frac{5}{8} + (^{-}1\frac{3}{4}) + 2 =$

12. $4.524 + 7.342 =$

13. ${}^{-}7\frac{2}{4} + 2\frac{3}{4} =$

14. $34 + (^{-}13) + 18 + 0 + 34 =$

15. $8.43 + (^{-}10.98) + (^{-}3.23) =$

16. $2.54 + (^{-}5.21) + (^{-}6.34) =$

17. ${}^{-}2\frac{1}{3} + (^{-}5\frac{7}{10}) + (^{-}7) =$

18. ${}^{-}1\frac{2}{3} + (^{-}3\frac{3}{5}) + 4 =$

19. $2\frac{1}{2} + 6\frac{1}{2} =$

20. $4\frac{3}{5} + (^{-}3\frac{2}{5}) + (^{-}8) =$

© Carson-Dellosa • CD-104632

Name _____ 7.NS.A.1c, 7.NS.A.1d

Subtracting Real Numbers

For all real numbers a and b, $a - b = a + (^{-}b)$. Simply stated: To subtract a number, add its opposite.

$7 - 10 = 7 + (^{-}10)$
$= ^{-}3$

$5 - 8 + 3 - 1 = 5 + (^{-}8) + 3 + (^{-}1)$
$= 5 + 3 + (^{-}8) + (^{-}1)$
$= 8 + (^{-}9)$
$= ^{-}1$

(group the numbers with like signs)

$^{-}4 - (^{-}12)$
$= ^{-}4 + 12$
$= 8$

Change each problem into an addition problem.

1. $7 - 9$

2. $^{-}6 - (^{-}4)$

3. $^{-}11 - 5$

4. $12 - (^{-}15)$

5. $8 - 3$

6. $22 - 5$

7. $4 - 11$

8. $^{-}4 - (^{-}9)$

Subtract.

9. $9 - 11$

10. $0 - (^{-}12)$

11. $^{-}5 - 4$

12. $6 - (^{-}6)$

13. $^{-}1 - (^{-}1)$

14. $^{-}7 - 6$

15. $3 - (^{-}5)$

16. $17 - 23$

© Carson-Dellosa • CD-104632

Name _____ 7.NS.A.1c, 7.NS.A.1d

Subtracting Real Numbers

$$10 - (^{-}4) = 10 + 4 = 14$$

Subtract.

1. $9 - (^{-}32) =$

2. $^{-}99 - (^{-}42) =$

3. $4 - (^{-}8) =$

4. $0 - 21 =$

5. $45 - 301 =$

6. $^{-}19 - 8 =$

7. $^{-}43 - 6 =$

8. $9 - (^{-}2) - 8 - 7 =$

9. $35 - 67 - 85 - 21 - 12 =$

10. $12 - 7 - (^{-}16) - 9 - (^{-}34) =$

11. $18 - (^{-}13) =$

12. $121 - 45 =$

13. $-\frac{4}{7} - \frac{1}{3} - (\frac{2}{3}) =$

14. $3.434 - 7.294 =$

15. $8 - 2.8 =$

16. $8 - (^{-}14) =$

17. $3.9 - 4.9 =$

18. $^{-}7 - (^{-}3) =$

19. $2.19 - 7.8 - 8.31 =$

20. $38 - 39 - (^{-}13) =$

© Carson-Dellosa • CD-104632

Name _____

7.NS.A.1c, 7.NS.A.1d

Subtracting Real Numbers

Subtract.

1. $^{-}9 - (^{-}5) =$

2. $321 - (^{-}34) =$

3. $\frac{2}{3} - \frac{4}{5} =$

4. $\frac{3}{5} - \frac{7}{8} =$

5. $5.34 - 9.9 - 3.65 =$

6. $9.432 + 4.348 - 44.938 =$

7. $245 - 32 - (^{-}36) =$

8. $44 - 35 - 34 - 32 =$

9. $8 - (^{-}5) - 7 - 9 =$

10. $43 - 88 - 35 - 21 =$

11. $-\frac{2}{5} - \frac{3}{4} - (-\frac{4}{5}) =$

12. $^{-}45 - 5 =$

13. $-\frac{2}{3} - \frac{1}{3} - (-\frac{1}{3}) =$

14. $-\frac{4}{5} - \frac{1}{2} - \frac{2}{5} =$

15. $4 - 12.9 =$

16. $7 - (^{-}33) =$

17. $3.4 - 7.4 =$

18. $2.456 - 4.345 - 5.457 =$

19. $23 - (^{-}21) =$

20. $4.346 - 0.4537 =$

© Carson-Dellosa • CD-104632

Name _____ 7.NS.A.2a, 7.NS.A.2c

Multiplying Real Numbers

The **property of zero for multiplication** states that for all real numbers a, $a \bullet 0 = 0$ and $0 \bullet a = 0$. Simply stated, any real number multiplied by 0 is 0. For example, $0 \bullet 20 = 0$.

To multiply two real numbers with the same signs:

1. Multiply their absolute values.
2. The sign of their product is positive.

positive • positive = positive
(+) (+) (+)
$3 \bullet 12 = 36$

negative • negative = positive
(−) (−) (+)
$-7 \bullet -8 = 56$

To multiply two real numbers with different signs:

1. Multiply their absolute values.
2. The sign of their product is negative.

negative • positive = negative
(−) (+) (−)
$-2 \bullet 5 = -10$

positive • negative = negative
(+) (−) (−)
$4 \bullet -8 = -32$

Write the sign of the product for each number.

1. $(-10)4$
2. $8(-1)$
3. $(-2)(-3)$
4. $(7)(5)(-3)$
5. $5(6)$
6. $(-2)(-7)$
7. $(-12)(-4)(-1)$
8. $(-6)(4)(-2)$

Multiply to find each product.

9. $4(6)(-1)$
10. $(-1)(-4)(-3)$
11. $(-\frac{1}{2})(2)$
12. $(7)(-3)(0)$
13. $(5)(3)$
14. $(-\frac{1}{8})(-16)(4)$
15. $(-9)(-4)$
16. $(-7)(7)$

© Carson-Dellosa • CD-104632

Name _____

7.NS.A.2a, 7.NS.A.2c

Multiplying Real Numbers

$(^{-}2)(^{-}3) = 6$

Multiply.

1. $4 \bullet 9 =$

2. $^{-}4 \bullet 12 =$

3. $(^{-}\frac{5}{9})(8.8) =$

4. $(^{-}3)(0) =$

5. $(^{-}3)(^{-}9) =$

6. $6(23) =$

7. $(12)(^{-}3)(4) =$

8. $(^{-}5)(^{-}5)(^{-}5) =$

9. $(5)(2)(^{-}1) =$

10. $(7)(^{-}9)(^{-}12) =$

11. $(^{-}\frac{2}{3})(^{-}1.6) =$

12. $^{-}7(^{-}7) =$

13. $(54.2)(^{-}3.55) =$

14. $(2.22)(^{-}1.11) =$

15. $(7.44)(3.2)(4.3) =$

16. $(2.4)(^{-}1.4) =$

17. $(^{-}\frac{3}{5})(\frac{3}{5}) =$

18. $(^{-}\frac{4}{5})(2.2) =$

19. $^{-}8 \bullet 12 =$

20. $(0)(2)(^{-}213) =$

© Carson-Dellosa • CD-104632

Name _____

7.NS.A.2a, 7.NS.A.2c

Multiplying Real Numbers

Substitute and multiply.

$x = {}^{-}3$, $y = {}^{-}5$, $z = 0$

1. xy
2. ${}^{-}3yz$
3. xyz
4. $4x({}^{-}2y)$
5. ${}^{-}3(10xy)$
6. $2y(5y)$

$x = {}^{-}2.4$, $y = 3.1$, $z = {}^{-}4.8$

7. $4xz$
8. $yz({}^{-}z)$
9. ${}^{-}5(xyz)$
10. ${}^{-}x({}^{-}6z)$
11. $(xy)({}^{-}3x)$
12. $(3yz)({}^{-}7x)$

$x = \frac{1}{2}$, $y = -\frac{2}{3}$, $z = \frac{3}{5}$

13. $8(xy)$
14. xyz
15. ${}^{-}12x({}^{-}5yz)$
16. $(xy)(3xz)$
17. $({}^{-}2z)(3y)(6x)$
18. ${}^{-}z(10xyz)$

© Carson-Dellosa • CD-104632

Name _____ 7.NS.A.2b, 7.NS.A.2c

Dividing Real Numbers

The **multiplicative inverse property** states that for each nonzero a, there is exactly one number $\frac{1}{a}$ such that: $a \bullet \frac{1}{a} = 1$ and $\frac{1}{a} \bullet a = 1$.

The number $\frac{1}{a}$ is called the reciprocal of a.

For example, the reciprocal of 9 is $\frac{1}{9}$ and the reciprocal of $\frac{1}{2}$ is 2. Simply flip the number to find its reciprocal.

To divide the number a by the number b, multiply a by the reciprocal of b.

$a \div b = a \bullet \frac{1}{b}$ (The result will be the quotient of a and b.)

$12 \div 4 = 12 \bullet \frac{1}{4}$
$= 3$

$4x \div \frac{1}{4} = 4x \bullet 4$
$= 16x$

$15 \div \frac{3}{2} = 15 \bullet \frac{2}{3}$
$= 10$

To divide two nonzero real numbers:

1. The quotient is positive if both numbers have the same sign.
2. The quotient is negative if both numbers have different signs.

$^-15 \div 5 = {}^-15 \bullet \frac{1}{5}$
$= {}^-3$

$3 \div \frac{1}{4} = 3 \bullet 4$
$= 12$

$1 \div -\frac{4}{5} = 1 \bullet -\frac{5}{4}$
$= -\frac{5}{4}$

Write the reciprocal of each number. Write none if it does not exist.

1. 2
2. 1
3. 0
4. $-\frac{1}{9}$
5. $-\frac{1}{4}$
6. $\frac{4}{3}$
7. $^-8$
8. 10

Divide.

9. $\frac{9}{3}$
10. $\frac{^-28}{7}$
11. $^-15 \div 3$
12. $\frac{^-7}{8} \div \frac{1}{8}$
13. $36 \div (^-4)$
14. $\frac{^-8}{^-8}$
15. $44 \div (^-11)$
16. $0 \div \frac{4}{5}$

© Carson-Dellosa • CD-104632

Name _____

7.NS.A.2b, 7.NS.A.2c

Dividing Real Numbers

$9 \div 4.5 = 2$

Divide.

1. $\frac{49}{7} =$

2. $90 \div 15 =$

3. $(^{-}12) \div (9.9) =$

4. $(-\frac{2}{3}) \div (^{-}18) =$

5. $^{-}42 \div 7 =$

6. $45 \div (^{-}8) =$

7. $\frac{^{-}36}{4} =$

8. $(-\frac{3}{5}) \div (\frac{3}{5}) =$

9. $^{-}72 \div (9) =$

10. $^{-}21 \div (^{-}9) =$

11. $\frac{102}{17} =$

12. $0 \div (^{-}8) =$

13. $\frac{95}{5} =$

14. $\frac{63}{^{-}9} =$

15. $(^{-}3.4) \div (^{-}9.99) =$

16. $^{-}50 \div [40 \div (^{-}20)] =$

17. $(-\frac{4}{6}) \div (36) =$

18. $56 \div (^{-}28 \div 7) =$

19. $(^{-}38 \div 19) \div (^{-}2) =$

20. $^{-}45 \div [^{-}20 \div (^{-}4)] =$

© Carson-Dellosa • CD-104632

Name _____

7.NS.A.2b, 7.NS.A.2c

Dividing Real Numbers

Substitute and divide.

$x = {}^{-}24$, $y = 6$, $z = {}^{-}8$

1. $\frac{x}{y}$

2. $\frac{yz}{{}^{-}x}$

3. $\frac{{}^{-}z}{xy}$

4. $\frac{4z}{5y}$

5. $\frac{{}^{-}xy}{{}^{-}yz}$

6. $\frac{2y}{{}^{-}5x}$

$w = 2.8$, $x = 1.4$, $y = {}^{-}8.4$, $z = {}^{-}11.2$

7. $(w) \div ({}^{-}x)$

8. $(z) \div ({}^{-}y)$

9. $\frac{x}{({}^{-}w)}$

10. $({}^{-}z) \div ({}^{-}w)$

11. $\frac{{}^{-}3y}{2x}$

12. $(z) \div ({}^{-}x)$

$x = {}^{-}\frac{1}{4}$, $y = {}^{-}\frac{1}{3}$, $z = \frac{4}{5}$

13. $(z) \div (x)$

14. $\frac{x}{{}^{-}z}$

15. $\frac{({}^{-}2x)}{(y)}$

16. $(y) \div ({}^{-}xz)$

17. $({}^{-}z) \div (y) \div ({}^{-}x)$

18. $\frac{{}^{-}z}{xy}$

© Carson-Dellosa • CD-104632

Name _____ 5.OA.A.1, 6.EE.A.2c

Order of Operations

When solving an equation, be sure to follow the **order of operations**.

1. Parentheses
2. Exponents
3. Multiplication & Division
4. Addition & Subtraction

$$14 - 54 \div 6 = 14 - 9 = 5$$

Solve.

1. $3 \times 15 \div 5 =$

2. $35 \div 5 - 9 =$

3. $3 + 2 \times 4 =$

4. $5 \times 2 \times 8 =$

5. $6 - 40 \div 8 =$

6. $12 - 30 \div 6 =$

7. $32 \div 4 \times 3 =$

8. $8 + 3 \times 2 =$

9. $4 + 12 \div 2 =$

10. $9 + 20 \div 5 =$

11. $15 - 75 \div 5 =$

12. $9 - 3 + 6 =$

13. $2 \times 8 \div 4 =$

14. $3 + 3 - 2 =$

© Carson-Dellosa • CD-104632

Name _____

5.OA.A.1, 6.EE.A.1, 6.EE.A.2c

Order of Operations

Remember to follow the **order of operations**.

1. Parentheses
2. Exponents
3. Multiplication & Division
4. Addition & Subtraction

$$(3^3 + 6 \times 5) - 2 = (27 + 6 \times 5) - 2 = (27 + 30) - 2 = 55$$

Solve.

1. $(3^2 + 2 \times 3) \div 5 =$

2. $5^2 - 4^2 + 2 =$

3. $(4 + 2)^2 =$

4. $(11 - 8)^3 =$

5. $2(7 + 2) =$

6. $(9 - 7)^3 - (4 + 3) =$

7. $(14 - 6)2 =$

8. $4 + 3(12 - 9) =$

9. $5^2 - 2^3 =$

10. $3 \times 8 - (3 \times 2 + 7) =$

11. $(5^2 - 3 \times 5) \div 2 =$

12. $7 + 2^2(5 + 2) =$

13. $3 + 7^2 =$

14. $(2^2 + 3)^2 - 4 =$

15. $6 + 7 \times 3 - 9 \times 2 =$

16. $(2 \times 3) + (21 \div 7) =$

17. $7^2 - 2(3 \times 3 + 5) =$

18. $3 + (6 \times 2) =$

© Carson-Dellosa • CD-104632

Name _____

5.OA.A.1, 6.EE.A.1, 6.EE.A.2c

Order of Operations

Solve.

1. $8 - 4 \bullet 5(3 - 2) + 3 =$

2. $12 \div (2 - 7) + 7 =$

3. $(14 - 9) + 4 =$

4. $\dfrac{3^2 - 5 \bullet 7 - 4^2}{(^-4 - 7 - 12) + 8} =$

5. $9(3 \div 3) + 4(^-5 \bullet 9) \div 3 =$

6. $3 - (6 \bullet 6) - 3 \bullet 0 =$

7. $36 \div 9 - 8 + 21 \div 3 =$

8. $5(3 - 8) \bullet 3 + 8 - 3 =$

9. $3 \bullet 5 + 9 \bullet 7 =$

10. $\dfrac{(5 - 9)^2 + 2}{(7 - 8)^2 \bullet 3^2} =$

11. $4^2 + 3^2 - 7^2 =$

12. $\dfrac{3^2 - 10}{4^2 - 12} =$

13. $8^2 - \dfrac{26}{(4 + 9)} + 4 =$

14. $\dfrac{5 \bullet 7 - (3 + 4)}{^-(2^2) - 2^2 + 3^2} =$

15. $\dfrac{4 + 2 \bullet 3 + 4 - 3}{2^2 \bullet 3^2 - 3} =$

16. $\dfrac{3 + 10 - 19 + 32}{3^2 - 1 + 2^2} =$

17. $12 \div [3 + (6 + 3)] =$

18. $3 \bullet (0 - 7) + 8 \div 2^2 =$

© Carson-Dellosa • CD-104632

Name _____

6.NS.C.5, 6.NS.C.6a, 6.NS.C.7

Real-Number Operations with Absolute Value

Every real number has an opposite. Opposite numbers are the same distance from 0 on a number line and lie on the opposite sides of 0. The opposite of a positive number is a negative number. The opposite of a negative number is a positive number. The numbers 3 and $^{-}3$ are opposites. Find their places on the real number line to the right. The additive inverse of a number is the same as the opposite of a number.

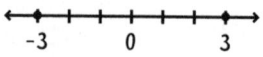

Remember, the opposite of 0 is simply 0 since it is neither positive nor negative.

The symbol $|x|$ is called the absolute value of x. The absolute value of a number is the distance between the number and 0 on a number line. The absolute value of a number, whether positive or negative, is always positive.

$|10| = 10$ $\quad$ $|^{-}2| = 2$ $\quad$ $|^{-}102| = 102$ $\quad$ $^{-}|12| = {}^{-}12$

Note: The answer to $^{-}|12|$ is $^{-}12$ because the absolute value of 12 is 12. But, it is multiplied by a negative, which resulted in $^{-}12$.

Write the opposite of each number.

1. 5
2. $^{-}2.6$
3. $^{-}40$
4. 2.8

Write a real number to represent each situation.

5. a gain of 12 yards
6. a temperature drop of 8°
7. a deposit of $89.26
8. a withdrawal of $75

Write the absolute value of each number.

9. $|10|$
10. $|23|$
11. $|0|$
12. $|^{-}42|$

Simplify.

13. $^{-}|^{-}15|$
14. $|^{-}5| \bullet |^{-}4|$
15. $^{-}|4 \bullet 3|$
16. $^{-}|30 \div 6|$
17. $|^{-}8| + |8|$
18. $|7| - |^{-}7|$

© Carson-Dellosa • CD-104632

Name _____ 6.NS.C.6a, 6.NS.C.7

Real-Number Operations with Absolute Value

The **absolute value** of a number is its distance from zero.

$-|5 - 11| = -|-6| = -6$ $|-4| \bullet |-3| = 4 \bullet 3 = 12$

Simplify.

1. $|-3| =$

2. $|-14| =$

3. $9 + |-4| =$

4. $-5 \bullet |4| + |5| =$

5. $-|4 \bullet 7| =$

6. $|21| + 9 =$

7. $23 + |8| =$

8. $|12| - |-15| =$

9. $7 - |-23| + |-7| =$

10. $|-6| \bullet |8| =$

11. $-|-3 + 7| =$

12. $-|-5 + 10| \div |-3| =$

13. $|-9| + |23| =$

14. $|-17| \div |-17| =$

15. $|1| - |0| + 6 =$

16. $|-67| - |-17| =$

17. $|4| - |-12 \div 4| =$

18. $|3 - 13| \div |-5| =$

19. $|24| \div |-12| =$

20. $-|9| \bullet |-9| =$

© Carson-Dellosa • CD-104632

Name _____

6.NS.C.6a, 6.NS.C.7

Real-Number Operations with Absolute Value

Simplify.

1. $|^{-}4 + 8| =$

2. $^{-}|5| \bullet |^{-}7| =$

3. $12 \div |^{-}3 + 7| =$

4. $^{-}|11| - |^{-}20| =$

5. $|6 - 18| + |^{-}17 + 9| =$

6. $|^{-}27| \div |^{-}3| \bullet {}^{-}|^{-}2| =$

7. $^{-}48 \div |^{-}2 \bullet 6| =$

8. $|^{-}83| - |^{-}38| =$

9. $|24| \bullet |^{-}4 + 1| =$

10. $10 - |^{-}13| + |^{-}7| =$

11. $|72 \div 8| \bullet |^{-}11| =$

12. $^{-}|16 - 21| - |^{-}7 + 4| =$

13. $23 + |^{-}45 \div 9| - |52| =$

14. $^{-}|^{-}4 - 15| =$

15. $|^{-}83| + |^{-}16| =$

16. $^{-}|4| \bullet |^{-}4| =$

17. $|^{-}7 + 15| \div |^{-}20 - 12| =$

18. $^{-}41 + |^{-}29| \bullet |0| =$

19. $|^{-}15| \div |3| - |45| =$

20. $|22 \div 11| \bullet |^{-}32 \div 8| =$

© Carson-Dellosa • CD-104632

Name _____ 6.EE.A.2b, 6.EE.A.3

Combining Like Terms

The expression $3x^2 + 2x + 1$ has three terms, $3x^2$, $2x$, and 1. A **term** is either a single number, a variable, or numbers and variables multiplied together. A term in an expression without a variable is called a **constant**, as 1 is above. For terms to be considered "like" terms, they must have the same variable and corresponding variables must have the same exponents. All constant terms are considered "like" terms.

like terms	unlike terms
$3x$ and $8x$	$9y$ and $10z$
$2x^2y$ and $3x^2y$	$3ab$ and $4ab^2$

In the example, $3x$ and $8x$ are like terms with numerical coefficients of 3 and 8. A **numerical coefficient** of a term is simply the number before its corresponding variable. When combining like terms, simply keep the variable the same and combine the numerical coefficients.

$4y + 10y = 14y$ $\qquad$ $6b + 9b - 5b = 10b$

$10x - 3x = 7x$ $\qquad$ $12x^2y - 10x^2y + 2x^2y = 4x^2y$

Identify the like terms in each problem.

1. $7c + 12c - 2$
2. $19y - 10$
3. $12rt - 10r + 18t$
4. $5r - 10r + 8rs$
5. $5d + 7d - 1$
6. $q + 9 + 2q + 5q$

Simplify. If not possible, write *already simplified.*

7. $8m - 3m$
8. $8y + 12y + 3y$
9. $3s + 8s - 2$
10. $2 + 10k$
11. $8q + 10q + 14$
12. $4 + 8x + 11y$
13. $5a + 6a - 9a$
14. $z + 8m + 4z - 4m$
15. $5w + 2 + w$

Name _____ 6.EE.A.2b, 6.EE.A.3

Combining Like Terms

$$4x + 5y + (^{-}18x) = {}^{-}14x + 5y$$

Combine like terms.

1. $3yz + 5yz =$

2. $3a + 5 + a =$

3. $5x - 5y - 8y + 8x =$

4. $18x + 3x =$

5. $5 - (^{-}4k) =$

6. $7c - 12c =$

7. $13ab + (^{-}12ab) =$

8. $^{-}12x + (^{-}4x) =$

9. $^{-}10n - (^{-}13n) =$

10. $12b + (^{-}34b) =$

11. $4.7x - 5.9x =$

12. $4x^2 + (^{-}8y) + (^{-}3xy) + 5x^2 + 2xy =$

13. $4x + 3y + (^{-}5y) + 3xy + y =$

14. $2x - y + 2x + 3xy =$

15. $5x + 7x =$

16. $23x + 8 + 6x + 3y =$

17. $^{-}e + 8e =$

18. $2xy + 5x + 6xy + 3xy + (^{-}3x) =$

19. $7s + 5x - 8s =$

20. $4xy + 7xy + 6x^2y + 3xy^2 =$

© Carson-Dellosa • CD-104632

Name _____

6.EE.A.2b, 6.EE.A.3

Combining Like Terms

Combine like terms.

1. $^{-}n + 9n + 3 - 8 - 8n =$

2. $3(^{-}4x + 5y) - 3x(2 + 4y) =$

3. $5 - 4y + x + 9y =$

4. $^{-}2x + 3y - 5x - {}^{-}8y + 9y =$

5. $6(a - b) - 5(2a + 4b) =$

6. $7(x + 5y) + 3(x + 5y) + 5(3x + 8y) =$

7. $12x + 6x + 9x - 3y + (^{-}7y) + y =$

8. $^{-}21x + (^{-}2x) =$

9. $4(x + 9y) - 2(2x + 4y) =$

10. $4(x + 5y) + (5x + y) =$

11. $6x + {}^{-}2y^2 + 4xy^2 + 3x^2 + 5xy^2 =$

12. $^{-}2(c - d) + (c - 3d) - 5(c - d) =$

13. $3x + (^{-}3y) - (4x) + y =$

14. $^{-}3(4x + {}^{-}2y) - 2(x + 3y) - 2(2x + 6y) =$

15. $2b + 3(2b + 8a) - 3(8b + 2a) =$

16. $3[2(^{-}y^2 + y)\,^{-}3] - 3(2x + y) =$

17. $2 \bullet 4x \bullet 3y - 4x \bullet 7y =$

18. $5(3a^2 - 2b^2) + 3a(a + 3b^2) =$

19. $3c + 4d + 2c + 5d - 4c =$

20. $4(x^2 + 3y^2) - y(x^2 + 5y) =$

© Carson-Dellosa • CD-104632

Name _____

8.EE.C.7b, HSA-REI.B.3

Solving One-Step Equations (Addition and Subtraction)

When solving equations for a given variable, use the addition or the subtraction property of equality. The addition property means that adding the same number to both sides of an equation will produce an equivalent equation.

$x - 3 = 5$ Solve for x by adding 3 to both sides of the equation.

$x - 3 + 3 = 5 + 3$

$x = 8$ When 3 is added to both sides, $^-3$ and 3 gives a result of 0, leaving x by itself. Therefore, $x = 8$ is the answer.

The subtraction property means that subtracting the same number from both sides of an equation will produce an equivalent equation.

$x + 3 = 5$ Solve for x by subtracting 3 from both sides of the equation.

$x + 3 - 3 = 5 - 3$

$x = 2$ When 3 is subtracted from both sides, $3 - 3$ gives a result of 0, leaving x by itself. Therefore, $x = 2$ is the answer.

Answers can be checked by substituting the value of x back into the equation. This is to make sure the value of x makes a true sentence when substituted into the original equation.

$x - 3 = 5, x = 8$
$8 - 3 = 5$
$5 = 5$

$x + 3 = 5, x = 2$
$2 + 3 = 5$
$5 = 5$

Solve each equation for x. Check your answers.

1. $x - 10 = 23$
2. $6 + x = {}^-11$
3. ${}^-13 = {}^-6 + x$
4. $7 = 14 + x$
5. $8 = x + 9$
6. $x - 5 = {}^-5$
7. $x + 3 = 12$
8. $7 + x = 7$
9. $7 + x = {}^-7$
10. ${}^-2 = x - 5$
11. $x + 4 = {}^-15$
12. $9 = x + 12$

© Carson-Dellosa • CD-104632

Name _____

8.EE.C.7b, HSA-REI.B.3

Solving One-Step Equations (Addition and Subtraction)

$$\begin{aligned} 12 + x &= {}^{-}24 \\ 12 - \ 12 + x &= {}^{-}24 - 12 \\ x &= {}^{-}36 \end{aligned}$$

Solve each equation for the given variable.

1. ${}^{-}13 + b = 31$

2. $x - 17 = {}^{-}27$

3. $27 = v + ({}^{-}5)$

4. ${}^{-}4 = x - 3$

5. $12 - ({}^{-}z) = 17$

6. ${}^{-}200 = b + ({}^{-}73)$

7. ${}^{-}13 + x = 18$

8. ${}^{-}w + ({}^{-}7) = {}^{-}56$

9. $3 + x = 9$

10. $z + 3.5 = 4.7$

11. $12 + ({}^{-}g) = 10$

12. $y - 12 = 15$

13. $x + 2 = {}^{-}6$

14. $s - 5 = {}^{-}8$

15. ${}^{-}13 = n + ({}^{-}39)$

16. $r = 4.4 + 3.9$

© Carson-Dellosa • CD-104632

Name _____ 8.EE.C.7b, HSA-REI.B.3

Solving One-Step Equations (Addition and Subtraction)

Solve each equation for the given variable.

1. $c + 23 = 41$

2. $48 - x = 15$

3. $a + 5.7 = 18.9$

4. $^{-}29 - n = 6$

5. $40 - (^{-}g) = 38$

6. $c - 3 = 4.7$

7. $n + \frac{3}{8} = \frac{5}{8}$

8. $^{-}1.9 + b = {}^{-}4.5$

9. $z + (^{-}14) = 13$

10. $^{-}135 + r = {}^{-}26$

11. $2\frac{1}{3} + w = 4\frac{2}{9}$

12. $^{-}0.6 - m = {}^{-}1.5$

13. $^{-}d + (^{-}61) = 107$

14. $s - \frac{7}{8} = -\frac{3}{4}$

15. $^{-}101 = g - (^{-}28)$

16. $241 + p = {}^{-}93$

17. $35.02 - q = 46.1$

18. $5\frac{3}{10} + (^{-}v) = 3.8$

19. $^{-}74 = {}^{-}k - (^{-}91)$

20. $^{-}8 + x = 5.62$

© Carson-Dellosa • CD-104632

Name _____

8.EE.C.7b, HSA-REI.B.3

Solving One-Step Equations (Multiplication and Division)

When solving equations for a given variable, use the multiplication or division property of equality. The multiplication property means that multiplying the same nonzero number by both sides of the equation will produce an equivalent equation.

$\frac{x}{7} = 3$ Multiply both sides of the equation by 7 to isolate x.

$(7)(\frac{x}{7}) = (7)(3)$

$x = 21$

Check $\frac{21}{7} = 3$

$3 = 3$

Therefore, the solution is 21.

The division property means that dividing both sides of the equation by the same nonzero number will produce an equivalent equation.

$^{-}5x = 20$ Divide both sides of the equation by $^{-}5$ to isolate x.

$\frac{^{-}5x}{^{-}5} = \frac{20}{^{-}5}$

$x = {}^{-}4$

Check $^{-}5(^{-}4) = 20$

$20 = 20$

Therefore, the solution is $^{-}4$.

Solve each equation for x. Check your answers.

1. $5x = 35$
2. $18 = {}^{-}3x$
3. ${}^{-}7x = 49$
4. $-\frac{1}{3}x = 6$
5. ${}^{-}5x = {}^{-}20$
6. $-\frac{5}{8}x = 10$
7. $\frac{1}{4}x = {}^{-}2$
8. $\frac{2}{3}x = 8$
9. $4 = -\frac{x}{5}$
10. ${}^{-}4x = 48$
11. $\frac{x}{3} = {}^{-}5$
12. ${}^{-}6x = 24$

Name _____ 8.EE.C.7b, HSA-REI.B.3

Solving One-Step Equations (Multiplication and Division)

$$3x = 15$$
$$\frac{3x}{3} = \frac{15}{3}$$
$$x = 5$$

$$-\frac{3}{4y} = -6$$
$$-\frac{4}{3} \bullet -\frac{3}{4y} = -6 \bullet -\frac{4}{3}$$
$$\frac{1}{y} = 8 \text{ so } y = \frac{1}{8}$$

Solve each equation for the given variable.

1. $-13h = 169$

2. $4b = -36$

3. $10x = -100$

4. $4c = 288$

5. $7x = -63$

6. $4y = -48$

7. $6x = -36$

8. $\frac{8}{k} = \frac{2}{5}$

9. $-(-90) = -45z$

10. $-50 = 2x$

11. $\frac{2}{n} = \frac{1}{9}$

12. $\frac{4}{x} = \frac{2}{9}$

13. $\frac{x}{6} = \frac{6}{9}$

14. $-35c = 700$

15. $-4x = -20$

16. $-\frac{x}{6} = \frac{2}{3}$

© Carson-Dellosa • CD-104632

Name _____

8.EE.C.7b, HSA-REI.B.3

Solving One-Step Equations (Multiplication and Division)

Solve each equation for the given variable.

1. $12.8 = 4b$

2. $-\frac{3}{4} = \frac{n}{16}$

3. $\frac{2}{3}x = {}^{-}10$

4. ${}^{-}({}^{-}36) = {}^{-}0.25c$

5. $1.6r = 80$

6. $-\frac{x}{8} = \frac{1}{4}$

7. $208x = {}^{-}4$

8. $\frac{2}{k} = \frac{1}{8}$

9. ${}^{-}4.9 = 7m$

10. $0.006w = {}^{-}0.54$

11. $\frac{720}{p} = 9$

12. $-\frac{5h}{50} = {}^{-}0.1$

13. ${}^{-}9s = 6$

14. $15d = -\frac{1}{3}$

15. ${}^{-}({}^{-}27)d = 16.2$

16. $3z = {}^{-}174$

17. $-\frac{4}{5j} = {}^{-}8$

18. $\frac{56}{63} = -\frac{x}{9}$

© Carson-Dellosa • CD-104632

Name _____ 8.EE.C.7b, HSA-REI.B.3

Solving Basic Multistep Equations

When solving equations for a given variable, sometimes you need to use more than one of the properties of equality.

$3x - 2 = 7$

$3x - 2 + 2 = 7 + 2$ Add 2 to both sides of the equation.

$\frac{3x}{3} = \frac{9}{3}$ Now, divide both sides of the equation by 3.

$x = 3$ Check $3(3) - 2 = 7$

$9 - 2 = 7$

$7 = 7$

Therefore, the solution is 3.

Here are some steps to follow when solving multi-step equations.

1. Simplify both sides of the equation (if possible).
2. Use the addition or subtraction property of equality to isolate terms containing the variable.
3. Use the multiplication or the division property of equality to further isolate the variable.
4. Check the solution.

Solve each equation for x. Check your answers.

1. $6x - 3 = 21$
2. $^{-}6 + \frac{x}{4} = 1$
3. $18 - 3x = {}^{-}12$
4. $7 + 2x = {}^{-}13$
5. ${}^{-}4 = 7x + 8 - 8x$
6. $13 = 9 - \frac{x}{5}$
7. ${}^{-}7 - x = {}^{-}5$
8. $5x + 9 - 4x = 12$
9. ${}^{-}8x - 13 = 19$
10. ${}^{-}3 = {}^{-}5 - 2x$
11. $\frac{1}{2}x + 9 = 15$
12. ${}^{-}7 = 3x - 15 - 7x$

© Carson-Dellosa • CD-104632

Name _____

8.EE.C.7b, HSA-REI.B.3

Solving Basic Multistep Equations

$$4x + 4 = 12$$
$$4x + 4 - 4 = 12 - 4$$
$$4x = 8$$
$$x = 2$$

Solve each equation for the given variable.

1. $7x - 12 = 2$

2. $7a - 4 = 24$

3. $4b - 7 = 37$

4. $3c - 9 = 9$

5. $8 - 9y = 35$

6. $8 - 12x = 32$

7. $1.3x + 5 = {}^{-}5.4$

8. $3(y + 4) + 5 = 35$

9. $0 = 25x + 75$

10. $3 - \frac{1}{5}e = {}^{-}7$

11. $5 - \frac{1}{2}x = {}^{-}9$

12. $2x = 6 + ({}^{-}18)$

13. $7 - \frac{1}{9}k = 32$

14. $\frac{3}{12}w + 2 = 11$

15. $\frac{2x}{5} + 3 = 9$

16. $\frac{x}{3} - 8 = {}^{-}12$

17. $5(e + 5) = {}^{-}10$

18. $8 - \frac{1}{2}y = {}^{-}6$

© Carson-Dellosa • CD-104632

Name _____

8.EE.C.7b, HSA-REI.B.3

Solving Basic Multistep Equations

Solve each equation for the given variable.

1. $5z - 8 = {}^{-}28$

2. $4k + 7 = {}^{-}9$

3. $13x + 7 = {}^{-}32$

4. $2x + 12 = 6$

5. $7.2 + 4x = 19.2$

6. $2(w - 6) = 8$

7. $7h + 1 = {}^{-}13$

8. $3(c - 2) = 15$

9. $6x - 5 = {}^{-}41$

10. ${}^{-}3 + 2n = {}^{-}15$

11. $5e + ({}^{-}9) = 26$

12. $\frac{m}{3} - 7 = {}^{-}10$

13. $6x - 2 = 34$

14. ${}^{-}8(r - 2) = 40$

15. $5n - 8 = {}^{-}23$

16. $2 + (\frac{1}{5})x = {}^{-}7$

17. $5 - (\frac{1}{2})g = 12$

18. $3x - 4 = 14$

19. ${}^{-}6 = \frac{3z}{4} + 12$

20. $2(f + 7) - 8 = 22$

21. $4.7 = {}^{-}3.4m - 5.5$

22. $32 = \frac{4}{6}x - 34$

© Carson-Dellosa • CD-104632

Name _____

8.EE.C.7b, HSA-REI.B.3

Solving Equations with Variables on Both Sides

Often, equations with a variable on both sides of the equation need to be solved. This requires one additional step of getting the variable on one side only.

$$4x - 1 = 2x + 7$$

$$4x - 2x - 1 = 2x - 2x + 7$$ Get the variable on one side of the equation.

$$2x - 1 = 7$$ Now, solve for x using the properties of equality.

$$2x - 1 + 1 = 7 + 1$$

$$\frac{2x}{2} = \frac{8}{2}$$

$$x = 4$$

Check $4(4) - 1 = 2(4) + 7$

$$16 - 1 = 8 + 7$$

$$15 = 15$$

Therefore, the solution is 4.

Combine the variables on the side of the equation with the greater variable coefficient, in order to avoid solving an equation with a variable which has a negative coefficient.

Solve each equation for x. Check your answers.

1. $9x - 12 = 3x$
2. $8x - 12 = 15x - 4x$
3. $11 + 6x = 2x - 13$
4. $^{-}5x = 9 - 2x$
5. $^{-}8x - 10 = 4x + 14$
6. $10x - 5 = 21 - 3x$
7. $^{-}12x = 14 - 5x$
8. $19 - 3x = 21 + x$
9. $4x + 12 = {}^{-}3x - 6 + 4x$
10. $14x + 5 = 8x - 1$
11. $7x - 3 = -4x - 25$
12. $^{-}9x + 5 = {}^{-}22 - 6x$

© Carson-Dellosa • CD-104632

Name _____

8.EE.C.7b, HSA-REI.B.3

Solving Equations with Variables on Both Sides

$$6x - 7 = x + 33$$
$$6x - x - 7 = x - x + 33$$
$$5x - 7 + 7 = 33 + 7$$
$$5x = 40$$
$$x = 8$$

Solve each equation for the given variable.

1. $4x - 6 = 8 - 3x$

2. $2(x + 3) = 12 - x$

3. $5(5 - z) = 4(z - 5)$

4. $4e + 6 = {}^{-}8 + 11e$

5. ${}^{-}4(x - 6) = 2(7 - 7x)$

6. $2m - 9 = 8m - 27$

7. $b + 9 = 6 + 2b$

8. ${}^{-}9j + 3 = {}^{-}32 - 4j$

9. $5(j - 4) = {}^{-}8 - j$

10. $8({}^{-}9x + 4) = {}^{-}3(6x + 9) + 5$

11. $3d - 5 = {}^{-}9 + 2d$

12. ${}^{-}2 + 6d = 34 - 3d$

13. $3(k + 4) = {}^{-}3 - 2k$

14. $5(m - 3) = 27 - 2m$

15. ${}^{-}j + 2 = {}^{-}14 - 5j$

16. ${}^{-}(x + 7) - 5 = 4(x + 3) - 6x$

17. $5(x - 1) = 2x + 4(x - 1)$

18. ${}^{-}3(5k + 5) = 54 - 12k$

19. $3e + 1 = 36 - 4e$

20. ${}^{-}6r + 12 = {}^{-}2 + 8r$

© Carson-Dellosa • CD-104632

Name _____

8.EE.C.7b, HSA-REI.B.3

Solving Equations with Variables on Both Sides

Solve each equation for the given variable.

1. $7 - 6a = 6 - 7a$

2. $3c - 12 = 14 + 5c$

3. $3x - 3 = {}^{-}3x + {}^{-}3$

4. $2x - 7 = 3x + 4$

5. $9a + 5 = 3a - 1$

6. $8(x - 3) + 8 = 5x - 22$

7. $5d + 7 = 4d - 9$

8. ${}^{-}10w + 6 = {}^{-}7w + {}^{-}9$

9. ${}^{-}7c + 9 = c + 1$

10. $2n + 6 = 5n - 9$

11. $\frac{5}{2}x + 3 = \frac{1}{2}x + 15$

12. $5 + 3x = 7(x + 3)$

13. $12m - 9 = 4m + 15$

14. $2(f - 4) + 8 = 3f - 8$

15. ${}^{-}6 - ({}^{-}2n) = 3n - 6 + 5$

16. $4(2y - 4) = 5y + 2$

17. $2(r - 4) = 5[r + ({}^{-}7)]$

18. $6(x - 9) = 4(x - 5)$

19. $4(z + 5) - 3 = 6z - 13$

20. $4e - 19 = {}^{-}3(e + 4)$

21. $4({}^{-}6x - 2) = {}^{-}22x$

22. ${}^{-}\frac{1}{3}x - 5 = 7 - x$

© Carson-Dellosa • CD-104632

Name _____

6.EE.A.2, 7.EE.B.3, 8.EE.C.7b

Writing and Solving Equations

The sum of three times a number and 25 is 40. Find the number.

$$3x + 25 = 40$$
$$3x + 25 - 25 = 40 - 25$$
$$3x = 15$$
$$x = 5$$

The number is 5.

Write an equation for each word problem and solve it.

1. The difference of a number and $^{-}3$ is 8. Find the number.

   Equation _____ **Solution** _____

2. Twice a number added to 9 is 15. Find the number.

   Equation _____ **Solution** _____

3. Twelve subtracted from 3 times a number is 15. Find the number.

   Equation _____ **Solution** _____

4. The sum of 4 times a number and 5 is $^{-}7$. Find the number.

   Equation _____ **Solution** _____

5. The product of a number and 5 is 60. Find the number.

   Equation _____ **Solution** _____

6. The difference of 5 times a number and 6 is 14. Find the number.

   Equation _____ **Solution** _____

7. The sum of a number and $^{-}6$ is 10. Find the number.

   Equation _____ **Solution** _____

8. The quotient of a number and 4 is $^{-}12$. Find the number.

   Equation _____ **Solution** _____

© Carson-Dellosa • CD-104632

Name _____ 6.EE.A.2, 7.EE.B.3, 8.EE.C.7b

Writing and Solving Equations

Write an equation for each word problem and solve it.

1. Six times the difference of a number and 9 is 54. Find the number.

   Equation _____ **Solution** _____

2. The sum of 8 times a number and 3 is 59. Find the number.

   Equation _____ **Solution** _____

3. The sum of 5 times a number and $^{-}11$ is $^{-}16$. Find the number.

   Equation _____ **Solution** _____

4. Twelve times the sum of a number and $^{-}8$ is 48. Find the number.

   Equation _____ **Solution** _____

5. The sum of 5 times a number and 2 is $^{-}13$. Find the number.

   Equation _____ **Solution** _____

6. The sum of 7 times a number and 11 is 81. Find the number.

   Equation _____ **Solution** _____

7. Three times the sum of a number and $^{-}2$ is $^{-}15$. Find the number.

   Equation _____ **Solution** _____

8. Five times the sum of a number and 2 is 35. Find the number.

   Equation _____ **Solution** _____

© Carson-Dellosa • CD-104632

Name _____ 6.EE.A.2, 7.EE.B.3, 8.EE.C.7b

Writing and Solving Equations

Write an equation for each word problem and solve it.

1. A number increased by 10 is 38. Find the number.

   Equation _____ **Solution** _____

2. A number times 8 is $^{-}96$. Find the number.

   Equation _____ **Solution** _____

3. Thirty-two is 7 less than 3 times a number. Find the number.

   Equation _____ **Solution** _____

4. A number multiplied by 4 and increased by 7 is $^{-}25$. Find the number.

   Equation _____ **Solution** _____

5. A number decreased by 12 is the same as 3 times the number. Find the number.

   Equation _____ **Solution** _____

6. Four times the sum of a number and 7 is 44 less than the number. Find the number.

   Equation _____ **Solution** _____

7. A number increased by 16 is the same as 8 times the sum of the number and 9. Find the number.

   Equation _____ **Solution** _____

8. Four times a number is decreased by 9 and then increased by 12. The result is 5 less than 2 times the number. Find the number.

   Equation _____ **Solution** _____

© Carson-Dellosa • CD-104632

Name _____

6.EE.B.8, HSA-REI.B.3

Solving Inequalities

$$^{-}10n + 5 \leq 55$$
$$^{-}10n + 5 - 5 \leq 55 - 5$$
$$^{-}10n \leq 50$$
$$n \geq {}^{-}5$$

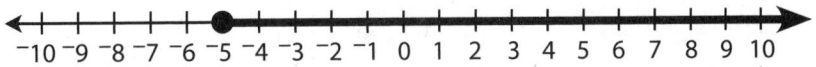

Solve each inequality and graph its solution set.

1. $^{-}4(3d + 2) \leq 4$

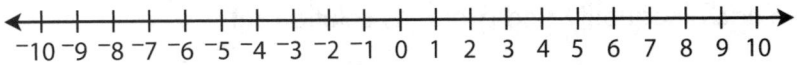

2. $10 - 5x - 20 \geq {}^{-}20$

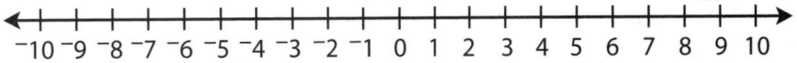

3. $^{-}15 > 4y - 7 - 3y - 4$

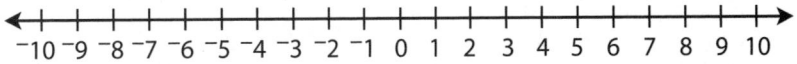

4. $4x - 7 < 9$

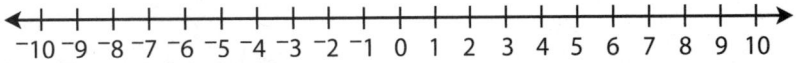

5. $6n - 3 > 33$

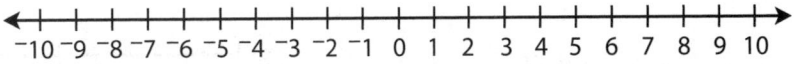

6. $3(3c - 4) \leq 15$

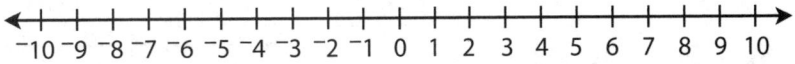

7. $5z - 1 > 9$

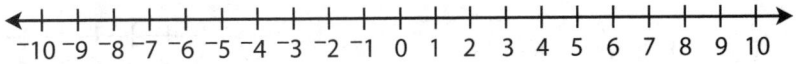

8. $5 > 4x - 11$

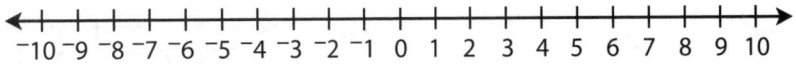

© Carson-Dellosa • CD-104632

Name _____

6.EE.B.8, HSA-REI.B.3

Solving Inequalities

$$^{-}10x > 4x - 42$$
$$^{-}10x + 10x > 4x + 10x - 42$$
$$0 > 14x - 42$$
$$0 + 42 > 14x - 42 + 42$$
$$42 > 14x$$
$$x < 3$$

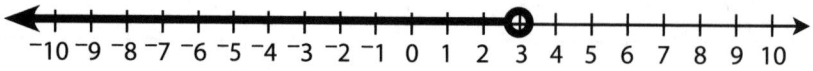

Solve each inequality and graph its solution set.

1. $^{-}2a < 10 + 3a$

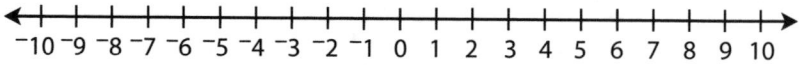

2. $4w > 2w + 8$

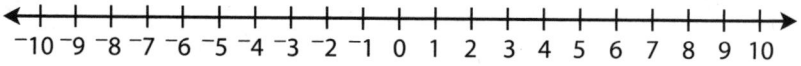

3. $2(k + 4) \leq 3(2k - 4)$

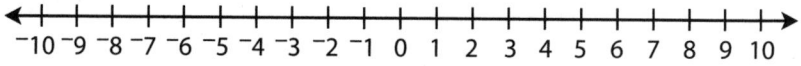

4. $5n + 3 \geq {}^{-}12$

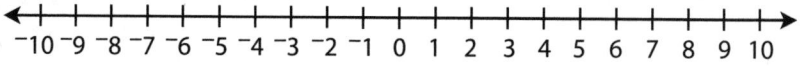

5. $5c + 2 < 2c + ({}^{-}7)$

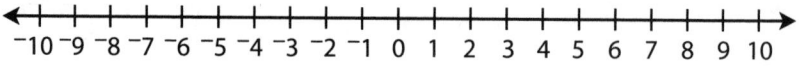

6. $5x - 20 > 2x + 1$

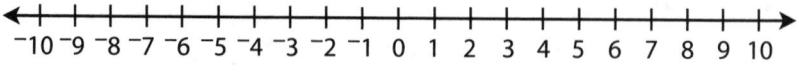

7. $3(s - 4) \geq 4s - 12$

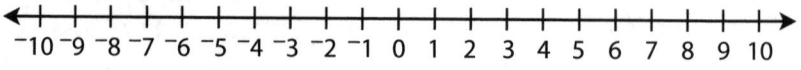

8. $^{-}9 - e > 3e + 11$

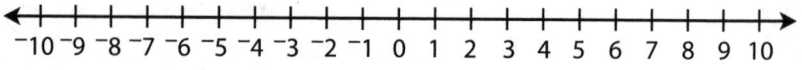

© Carson-Dellosa • CD-104632

Name _____

6.EE.B.8, HSA-REI.B.3

Solving Inequalities

Solve each inequality and graph its solution set.

1. $15e - 3 \leq 20e + 17$

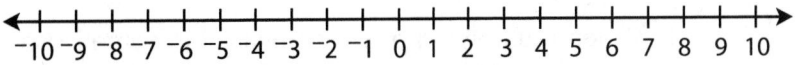

2. $6x - 4 > 2(x - 6)$

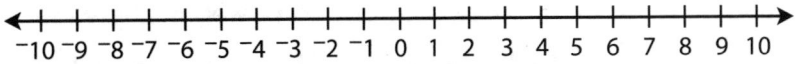

3. $7c - 8 \geq 6$

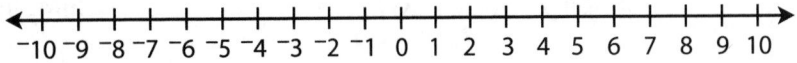

4. $5x + (^{-}3) \geq 2(3 + x)$

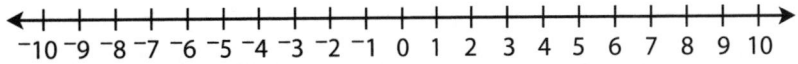

5. $^{-}8 < 2(2 + 3r)$

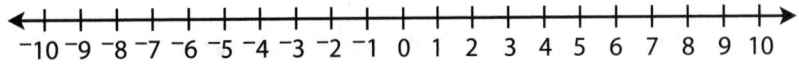

6. $6d < 3d - 18$

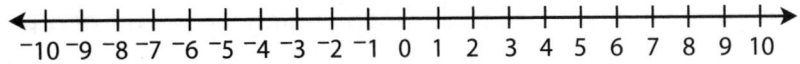

7. $7m + 9 \leq 6(m + 3)$

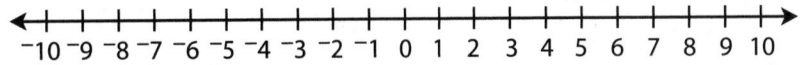

8. $3(2x + 4) \geq 7x + 8$

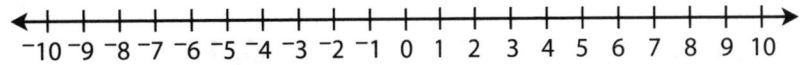

9. $^{-}4.2 > 0.6y$

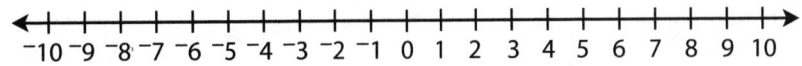

10. $\frac{a}{4} + 3 \leq 5$

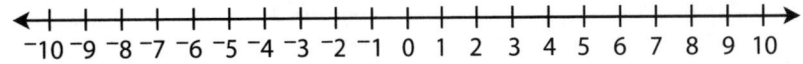

© Carson-Dellosa • CD-104632

Name _____

8.EE.A.1

Multiplication and Powers of Exponents

When multiplying exponents, it is important to remember the following properties:

1. When multiplying powers having the same base, add the exponents, keeping the same base. (Remember: in a^3, a is the base, 3 is the exponent, and a^3 is the power.) For example, $x^3 \bullet x^5 = x^{3+5} = x^8$.
2. When finding a power of a power, multiply the exponents. For example, $(x^3)^2 = x^6$.
3. When finding the power of a product, find the power of each factor and multiply. For example, $(x \bullet y)^2 = x^2 \bullet y^2$.

Simplify $(5x^3)^2(xy)^3$
$(5^2x^6)(x^3y^3)$
$= 25x^9y^3$

Simplify $(2x^4)^3(^-x^2)^3$
$(2^3x^{12})(^-x^6)$
$= {}^-8x^{18}$

1. What do you do with the exponents when multiplying powers that have the same base?

2. Label the base, the exponent, and the power in x^3.

3. Explain what you are to do when finding the power of a product.

Simplify each expression.

4. $3x \bullet x^2$
5. $(^-6xy)^2(x^2y)^3$
6. $(8xy)^2$
7. $^-4x^4 \bullet x^3$
8. $(^-3x^2y)^3$
9. $(4c^2)(^-5c^7)$
10. $(^-x^2)(^-x)^2$
11. $^-xy(^-xy)^2$
12. $(3x^3)(5x^5)$
13. $(x^3y^3)^3$

© Carson-Dellosa • CD-104632

Name _____ 8.EE.A.1

Multiplication and Powers of Exponents

Rule: $(x^a)^b = x^{ab}$ Example: $(x^2y^3)^3 = x^6y^9$

Rule: $x^a \bullet x^b = x^{a+b}$ Example: $x^3 \bullet x^5 = x^8$

Simplify each expression.

1. $c \bullet c^2 \bullet c^3 =$

2. $e \bullet e^2 \bullet e^3 \bullet e^4 \bullet e^5 =$

3. $a^3 \bullet a^4 \bullet a^7 \bullet a =$

4. $(3xy^2)(2x^2y^3) =$

5. $(2a^2b)(4ab^2) =$

6. $(5f)(^-3f^3)(2f)^2 =$

7. $(m^2n)^3(4mn^2)(mn) =$

8. $(4k^2)(^-3k)(3k^5) =$

9. $(^-2c^4)(4cd)(^-cd^2) =$

10. $(3x^3)^2(3x^4)(^-3x^2)^3 =$

11. $(^-1)(x)(^-x^2)(x^3)(^-x^2) =$

12. $(3x^2)(^-3x^5) =$

13. $(c^2h)^2(3ch^3)(2c^3h^4)^2 =$

14. $(^-4p^3)(^-4p^6)(^-2p^9) =$

15. $(12c^3)(2g^3)(4ch)^3 =$

16. $(4x^2y^3)^3(x^3y)(^-x^2y^2) =$

17. $(^-4f^3)^4(^-3m^3)^5 =$

18. $(2c^2d^2)^4(^-5cd^4) =$

19. $(3x)^2(2x^3y^6)(^-5x^6y^2) =$

20. $(3x)(^-4y^2)(6x^3y) =$

© Carson-Dellosa • CD-104632

Name _____

8.EE.A.1

Multiplication and Powers of Exponents

Simplify each expression.

1. $(^{-}4xy^3)^3 =$

2. $(x^2y^3)(x^3y) =$

3. $(^{-}6x^4y^6)^3 =$

4. $(5x^2y^4)^3 =$

5. $(6x^5y^4)(5x^5y^4) =$

6. $(2x)^4(x^3y)^2 =$

7. $(4x^3y^2)^3(^{-}2x^2y^4) =$

8. $(3x^2y)(^{-}8xy^4) =$

9. $(^{-}2x^2y)^4 =$

10. $(x^2y^3)^2(x^3y^2)^4 =$

11. $(^{-}4x^3y^3)^4(^{-}8x^3)^2 =$

12. $(3xy^3)(^{-}4x^2y^4)^2(xy^3) =$

13. $(^{-}3x^3y)^3 =$

14. $(^{-}2x^4y^5)^3 =$

15. $(3x^2y^3)^4(x^3y)(7xy^3) =$

16. $(6x^2y^3)(4xy^2)^3(3x^2y)^2 =$

17. $(^{-}2x^3y^3z)^4(2xyz^4)^2 =$

18. $(5xy^3)(^{-}5xy^2)^5 =$

19. $(^{-}3x^2y^3)^2 =$

20. $(4z^4)^2(2x^2y)(^{-}3xy^3z^5) =$

© Carson-Dellosa • CD-104632

Name _____ 8.EE.A.1

Division of Exponents

When dividing exponents, it is important to remember the following properties:

1. When dividing powers that have the same base, subtract the exponents.

   For example, $\frac{x^4}{x^2} = x^{4-2} = x^2$, where x cannot be equal to 0.

2. When finding a power of a quotient, find the power of the numerator and the power of the denominator and divide.

   For example, $(\frac{x}{y})^3 = \frac{x^3}{y^3}$, where y cannot be equal to 0.

Simplify:

$\frac{6^8}{6^6}$

$= 6^{8-6}$

$6^2 = 36$

$(\frac{3}{4})^{-2}$

$= \frac{3^{-2}}{4^{-2}}$

$= \frac{4^2}{3^2} = \frac{16}{9}$

1. Explain what you do with the exponents when dividing powers that have the same base.

Evaluate each expression.

2. $\frac{5^6}{5^3}$
3. $\frac{(-3)^2}{3^2}$
4. $\frac{3^5}{3^2}$
5. $\frac{5^4 \bullet 5}{5^7}$
6. $\frac{8^6}{8^3}$
7. $\frac{7^3}{7}$
8. $\frac{4^8}{4^8}$
9. $\frac{6^4 \bullet 6^3}{6^5}$
10. $\frac{4^6}{4^2}$
11. $\frac{10^5}{10^3}$

Simplify each expression.

12. $\frac{x^5}{x^3}$
13. $x^7 \bullet \frac{1}{x^5}$
14. $\frac{18x^4y^4}{6x^2y^4}$
15. $\frac{4x^4y^4}{4x^2y}$
16. $\frac{6x^2y^4}{3y^2}$
17. $x^4 \bullet \frac{1}{x^2}$

© Carson-Dellosa • CD-104632

Name _____

8.EE.A.1

Division of Exponents

Rule: $\frac{x^a}{x^b} = x^{a-b}$ Example: $\frac{x^6}{x^4} = x^{6-4} = x^2$ $\frac{x^3}{x^{-2}} = x^{3-(-2)} = x^5$

Simplify each expression.

1. $\frac{-12m^5}{6m} =$

2. $\frac{x^3}{x^2} =$

3. $\frac{9a^3b^5}{-3ab^2} =$

4. $\frac{16c^3}{-4c^2} =$

5. $\frac{d^3}{d^2} =$

6. $\frac{-3p^8}{6p^2} =$

7. $\frac{-54c^2d^4}{-8cd} =$

8. $\frac{49r^{13}}{-7r^8} =$

9. $\frac{45k^7r^3}{-3k^5} =$

10. $\frac{-14c^{15}d^3}{-2c^9d} =$

11. $\frac{(5k)(-8k^5)}{10k^3} =$

12. $\frac{24x^2y}{-4x^2} =$

13. $\frac{4x^2y^3z^4}{2xy^2z^3} =$

14. $\frac{9a^{11}}{a^3} =$

15. $\frac{22y^5z^8}{2yz^7} =$

16. $\frac{b^{14}c^9}{b^5c^4} =$

© Carson-Dellosa • CD-104632

Name _____ 8.EE.A.1

Division of Exponents

Simplify.

1. $\frac{a^5}{a^3} =$

2. $\frac{a^5b^2}{2a^2} =$

3. $\frac{13m^6n^7}{39m^3n^5} =$

4. $\frac{9x^8y^7z^8}{18x^5y^5z^4} =$

5. $\frac{^-2c^2d^3}{^-8cd^2} =$

6. $\frac{10a^6b^8}{40a^2b^2} =$

7. $\frac{18a^6b^2c^6}{36a^4bc^2} =$

8. $\frac{5x^3y^2z^2}{5x^2yz} =$

9. $\frac{45x^9y^{10}z^5}{51x^9y^8z^3} =$

10. $\frac{16x^2y^4}{4x^2y^3} =$

11. $\frac{18x^6y^3z^4}{12x^3y^2z^3} =$

12. $\frac{72x^5y^5z^6}{8x^4yz^3} =$

13. $\frac{18a^9b^3}{12a^2b^2} =$

14. $\frac{44x^8y^2}{11x^7y} =$

15. $\frac{(6x^3)(3x^8)}{^-12x^{10}} =$

16. $\frac{21k^9}{(3k)(7k^4)} =$

17. $\frac{(110c^3)(^-c^9)}{11c^5} =$

18. $\frac{(3xy)(4x^2y)}{^-6xy^2} =$

© Carson-Dellosa • CD-104632

Name _____

8.EE.A.1

Negative and Zero Exponents

Given a nonzero number a and a positive integer n, the following definitions of negative exponents and zero exponents are stated.

1. For a negative exponent: the expression a^{-n} is the reciprocal of a^n.
   This is written: $a^{-n} = \frac{1}{a^n}$, where $a \neq 0$.
2. For a zero exponent: any nonzero number raised to the 0 power will have an answer of 1. This is written: $a^0 = 1$, where $a \neq 0$.

Evaluate: 5^{-3}

$= \frac{1}{5^3}$

$= \frac{1}{125}$

Simplify by rewriting with positive exponents. $2x^{-3}y^2z^{-4}$

$= 2 \bullet \frac{1}{x^3} \bullet y^2 \bullet \frac{1}{z^4}$

$= \frac{2y^2}{x^3z^4}$

Note: When rewriting expressions to positive exponent form, simply move the factors from the denominator to the numerator or vice versa, leaving out the in-between step.

1. Any number raised to the zero power has what value?

Evaluate each expression.

2. 4^{-2}
3. $7^3 \bullet 7^{-3}$
4. $3 \bullet 3^{-1}$
5. $-5^0 \bullet \frac{1}{3^{-3}}$
6. $(6^2)^{-2}$
7. $(-2^{-3})^{-1}$

Rewrite each expression with positive exponents.

8. x^{-8}
9. $\frac{1}{3x^{-2}}$
10. $(-3)^0x^{-2}$
11. x^{-10}
12. $x^{-3}y^4$
13. $\frac{6}{x^{-2}}$

© Carson-Dellosa • CD-104632

Name _____ 8.EE.A.1

Negative and Zero Exponents

Rule: $x^{-a} = \frac{1}{x^a}$ Example: $4^{-2} = \frac{1}{16}$ Example: $4x^{-2} = \frac{4}{x^2}$ Example: $(2x)^{-3} = \frac{1}{8x^3}$

$4^{-2} = \frac{1}{4^2} = \frac{1}{16}$ $\frac{1}{(2x)^3} = \frac{1}{8x^3}$

Rule: $x^0 = 1$

Simplify.

1. $4cd^{-5}$
2. 3^0
3. $3a^4b^{-3}$
4. 4^{-5}
5. $(^-2)^{-2}$
6. $(3xy)^{-1}$
7. $(3x)^{-3}$
8. $7x^{-3}$
9. $^-2x^{-3}$
10. $(6y^2)^{-2}$
11. $(\frac{4}{5})^{-2}$
12. $4m^3n^{-5}$
13. $(^-11x^3y)^{-2}$
14. $(c^0d)^{-2}$
15. $(\frac{x^2}{y^3})^{-2}$
16. $(\frac{2}{3})^{-1}$
17. b^0
18. c^{-7}

© Carson-Dellosa • CD-104632

Name _____

8.EE.A.1

Negative and Zero Exponents

Simplify.

1. $6x^{-3}$

2. $4x^{-5}$

3. $\frac{1}{5x^{-4}}$

4. $4x^{-4}y^{-2}$

5. $\frac{1}{(2x)^{-3}}$

6. $(3x^{-2})^2$

7. $14x^{-8}y$

8. $(^{-}5x^3)^{-2}$

9. $(^{-}3x^{-2}y^5z^3)^{-2}$

10. $8x^{-2}y^{-4}$

11. $2x^0y^{-5}z^2$

12. $(4x^3y^{-4})^2$

13. $(\frac{x^3}{2y^{-2}})^{-3}$

14. $4m^{-5}n^{-1}$

15. $(\frac{x^2}{y^3})^{-2}$

16. $(^{-}14c^2d)^{-1}$

17. $9xy^{-4}$

18. $(^{-}6s^3p^2)^{-3}$

© Carson-Dellosa • CD-104632

Name _____ 7.EE.A.1, HSA-APR.A.1

Adding and Subtracting Polynomials

When adding polynomials, the following two methods can be used.

1. The horizontal method:

   Add $(3x^4 + 8x^2 - 2)$ and $(^-4x^2 + 7x^4 + 5)$.
   First group like terms in descending order. $(3x^4 + 7x^4) + (8x^2 - 4x^2) + (^-2 + 5)$
   Simplify. $10x^4 + 4x^2 + 3$

2. The vertical method:

   Add $(4x^7 + 6x^4 - 10 + 2x^5)$ and $(10x^5 - 5x^7 - 3x^4 + 12)$.
   Group like terms in columns in descending order.

$$\begin{array}{rr} & 4x^7 + 2x^5 + 6x^4 - 10 \\ + & {}^-5x^7 + 10x^5 - 3x^4 + 12 \\ \hline & {}^-x^7 + 12x^5 + 3x^4 + 2 \end{array}$$

   Simplify.

When subtracting polynomials, change the sign of each term in the second polynomial and then choose from one of the methods above to simplify.

Subtract. $(10x^7 - 4x^4 + 7x^2 + 2) - (12x^4 + 6x^7 + 8 - 5x^2)$

Change sign of each term in second polynomial. $10x^7 - 4x^4 + 7x^2 + 2 - 12x^4 - 6x^7 - 8 + 5x^2$

Choosing method 1 above, group like terms in descending order.

$(10x^7 - 6x^7) + (^-4x^4 - 12x^4) + (7x^2 + 5x^2) + (2 - 8)$

Simplify. $4x^7 - 16x^4 + 12x^2 - 6$

1. What is important to remember to do when subtracting polynomials?

Simplify.

2. $(7x - 5) - (3x + 7)$

3. $(2x^2 - 3x + 1) - (5x^2 - 3x + 10)$

4. $(6x + 7) + (8x - 3)$

5. $(9x^2 + 3x - 5) - (3x^2 + 4x - 10)$

6. $(8x^4 - 9x^3 + 2x) - (6x^4 + 7x^3 - 3x)$

7. $(x^3 - x^2 + 2) - (x^3 + x^2 + 5)$

© Carson-Dellosa • CD-104632

Name _____

7.EE.A.1, HSA-APR.A.1

Adding and Subtracting Polynomials

$$(x^2 + 4x + 2) - (2x^2 + 7x - 6) = {}^{-}x^2 - 3x + 8$$

Add or subtract the polynomials by combining like terms.

1. $(4y^2 - 9y) - ({}^{-}5y^2 + 8y - 8) =$

2. $(6x^2 + 2x + 6) - (4x^2 - 2x + 3) + ({}^{-}5x^2 + 5x + 6) =$

3. $({}^{-}2x^3 + 3x^2 + 9) + ({}^{-}8x^3 - 2x^2 + {}^{-}4x) =$

4. $(2x^2 - 9x - 8) - (2x^3 - 7x^2 + {}^{-}2) =$

5. $(4x^3 - 2x^2 - 12) + (6x^2 + 3x + 8) =$

6. $(3x^4 - 3x + 1) - (4x^3 - 4x - 8) =$

7. $(7x^2 - x - 5) - (3x^2 - 3x + 5) =$

8. $(x^3 - x^2 + 3) - (3x^3 - x^2 + 7) =$

9. $({}^{-}2x^2 + 4x - 12) + (5x^2 - 5x) =$

10. $(4x^3 - 5x^2 - 9) - (6x^3 - 5x - 4) - (5x^3 - 4x^2 - 10) =$

© Carson-Dellosa • CD-104632

Name _____

7.EE.A.1, HSA-APR.A.1

Adding and Subtracting Polynomials

Add or subtract the polynomials by combining like terms.

1. $(x^3 - 7x^4 + 2x^2 - 3x) + (4x^4 - 8x^3 + 5x) =$

2. $(6x^2 + 9x + 4) - (^-7x^2 + 2x - 1) =$

3. $(^-8x^2 + 2) + (7x - 5) =$

4. $(8x + 6) - (10x + 16) + (4x - 6) =$

5. $(6x^2 + 7x - 9) + (3x + 8) =$

6. $(4x^2 + 2) + (3x^2 - 4x + 6) - (5x^2 + 10) =$

7. $(2x^2 + 3x + 2) - (6x^3 - 3x^2 + 8) - (^-2x^3 + 9x^2 + 7) =$

8. $(^-3x^2 - 4x^3 - 1) - (2x^3 - 7x - 9) - (2x^3 - 2x^2 - 3) =$

9. $(^-6x^2 - 3x^3 + 4) + (^-7x^3 + 2x + 4) - (^-3x^3 + 5x^2 + 2) =$

10. $(4x^2 + 6x + 3) + (3x^2 - 3x - 2) + (^-4x^2 + 3x - 9) =$

11. $(9x^2 - 7x + ^-4) + (3x^3 - 4x + ^-5) + (^-4x^2 - 2x - 5) =$

12. $(3x^3 - 4x + 3) + (^-3x^3 + 5x - 8) =$

13. $(^-5x^5 + 7x^3 - 4x + 2) + (7x^5 - 8x^4 + 9x^3) =$

14. $(4x^2 - 8x + 2) - (^-5x^2 + 5x + 6) =$

15. $(10x^3 - 8x^2 + 6x - 1) - (^-7x^3 + 2x - 4) =$

© Carson-Dellosa • CD-104632

Name _____

6.EE.A.3, 7.EE.A.1, HSA-APR.A.1

Multiplying Polynomials and Monomials

To multiply a multi-termed polynomial by a monomial, simply multiply each term in the polynomial by the monomial using the distributive property. Remember the properties of exponents when simplifying such problems.

Multiply.	$3x(3x^3 - 2x^2 + 8x - 7)$	
	$3x(3x^3) - 3x(2x^2) + 3x(8x) - 3x(7)$	Distribute $3x$ to each term.
	$9x^4 - 6x^3 + 24x^2 - 21x$	Multiply to simplify.

You may have to simplify an expression using the distributive property to multiply the polynomial by a monomial.

Simplify.	$4x^7 + 9x^4 - 7 + 2x^2(8x^5 - 5x^2)$	
	$4x^7 + 9x^4 - 7 + 2x^2(8x^5) - 2x^2(5x^2)$	Distribute the $2x$.
	$4x^7 + 9x^4 - 7 + 16x^7 - 10x^4$	Multiply to begin simplifying.
	$20x^7 - x^4 - 7$	Simplify.

Multiply each expression.

1. $6x(4x - 3)$
2. $4x(x^2 - 6x + 3)$
3. $(^-3x^2 + 4x - 2)(^-4x^3)$
4. $x(^-5x + 2)$
5. $^-3x(5x^2 - 2x - 6)$
6. $(^-x^2 + 2x - 1)(5x^3)$
7. $^-2x(^-4x - 8)$
8. $^-x^2(7x^2 - x + 4)$
9. $^-5x^4(2x^2 - 8x + 6)$
10. $5y(y^2 - 3y + 1)$
11. $(3x^2 - 4x)(^-x)$
12. $7x(x^3 - x^2)$

© Carson-Dellosa • CD-104632

Name _____

6.EE.A.3, 7.EE.A.1, HSA-APR.A.1

Multiplying Polynomials and Monomials

$$4y(y - 3) = 4y^2 - 12y$$

Use the distributive property to multiply the polynomials.

1. $a(a + 8) =$

2. $5b(4b^3 - 6b^2 - 6) =$

3. $3x(x - 3) =$

4. $4a(2a + 6) =$

5. $y(y - 7) =$

6. $^-2x^2(5 - 3x + 3x^2 + 4x^3) =$

7. $4b(3 - b) =$

8. $2xy(2x - 3y) =$

9. $^-5y^2(7y - 8y^2) =$

10. $4x^2(3x^2 - x) =$

11. $x(x^2 + x + x) =$

12. $3b(4b^3 - 12b^2 - 7) =$

13. $(^-7x^3)(3x^2 - 1) =$

14. $^-5ab(6a - 4b) =$

15. $3x(x - 3) =$

16. $^-3x^2(4x^2 - 3x + 3) =$

17. $^-4x^2(3x^3 + 8x^2 + {}^-9x) =$

18. $(3x^4 - 5x^2 - 4)(^-3x^3) =$

© Carson-Dellosa • CD-104632

Name _____

6.EE.A.3, 7.EE.A.1, HSA-APR.A.1

Multiplying Polynomials and Monomials

Simplify each expression.

1. $^{-}4x^2 - 5x + 7 + 3(x^2 + 8x - 2)$

2. $5x^2 - 3x(x - 7)$

3. $6x^2 + 4x + (7x - 3)2x$

4. $^{-}x^2 + 8x - 6 - 7(3x^2 - 5x + 9)$

5. $6x^2 + 3(x - 5) - 8x$

6. $2x^2 + 7x + 6 - 5x(^{-}2x - 1)$

7. $^{-}3x^2 - 3x(4x^2 - 5x + 7) - 8x^2$

8. $x^3 + (7x^2 - 9x - 1)x + 10x^2$

Write and simplify an expression to find the area of each polygon. Remember that the area of a triangle can be found using $A = \frac{1}{2}bh$ and the area of a rectangle can be found using $A = lw$.

9.

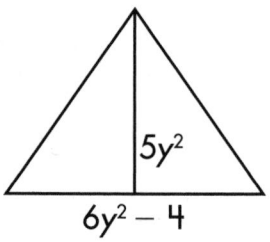

A = _____

10.

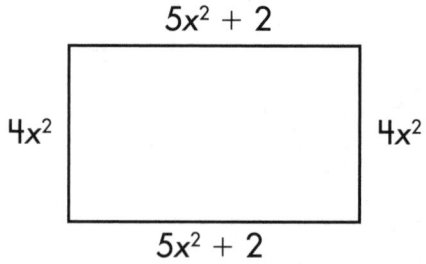

A = _____

11.

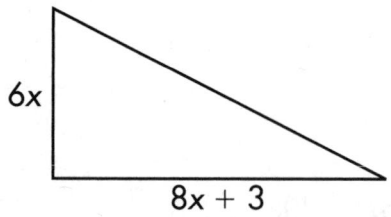

A = _____

12.

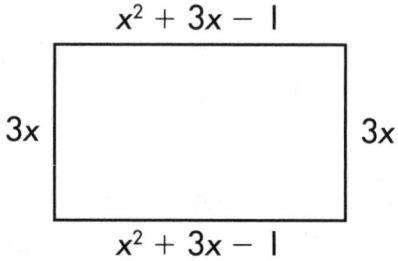

A = _____

© Carson-Dellosa • CD-104632

Name _____ 7.EE.A.1, HSA-SSE.A.2, HSA-APR.A.1, HSA-APR.C.4

Multiplying Binomials

The easiest way to multiply binomials is to use what is called the FOIL method. This method multiplies the first terms of the binomials (F), multiplies the outer terms of the binomials (O), multiplies the inner terms of the binomials (I), multiplies the last terms of the binomials (L), and simplifies by adding like terms.

1. Multiply $(3x - 2)(4x + 3)$

$3x(4x) = 12x^2$	Multiply first terms.	F
$3x(3) = 9x$	Multiply outer terms.	O
$^-2(4x) = {}^-8x$	Multiply inner terms.	I
$^-2(3) = {}^-6$	Multiply last terms.	L
$12x^2 + 9x - 8x - 6$	Simplify by combining like terms.	
$12x^2 + x - 6$	Final product	

2. Multiply $(5x + 2)(7x - 3)$

$5x(7x) + 5x(^-3) + 2(7x) + 2(^-3)$

F O I L

$35x^2 - 15x + 14x - 6$	Multiply.
$35x^2 - x - 6$	Final product

Multiply.

1. $(x + 7)(x - 5)$
2. $(x - 5)(3x + 7)$
3. $(9x - 1)(6x + 2)$
4. $(7x - 8)(8x - 7)$
5. $(x + 2)(x + 3)$
6. $(2b - 8)(3b - 7)$
7. $(3x + 5)(^-4x - 7)$
8. $(^-3x - 9)(^-x - 6)$
9. $(x - 10)(x + 1)$
10. $(2x - 3)(5x + 4)$
11. $(^-5x + 6)(x - 2)$
12. $(^-4x + 4)(5x + 8)$
13. $(5b - 2)(3b + 2)$
14. $(3x + y)(3x - 2y)$
15. $(4a + 1)(3a - 3)$
16. $(x - y)(2x + 2y)$

Name _____ 7.EE.A.1, HSA-SSE.A.2, HSA-APR.A.1, HSA-APR.C.4

Multiplying Binomials

Rules: $(a + b)(a - b) = a^2 - b^2$
$(a + b)^2 = a^2 + 2ab + b^2$
$(a - b)^2 = a^2 - 2ab + b^2$

Multiply.

1. $(2x + y)(2x - y) =$

2. $(b - 5)(b + 5) =$

3. $(x - y)(2x + 2y) =$

4. $(x - 4y)^2 =$

5. $(7x - 3y)(7x + 3y) =$

6. $(x + 6)(x - 6) =$

7. $(7x + y)(7x - y) =$

8. $(2x - 6y)^2 =$

9. $(3a - b)(3a + b) =$

10. $(x + 2)(x - 2) =$

11. $(12b - 5)(12b + 5) =$

12. $(2x - 3y)^2 =$

13. $(7x - 5y)^2 =$

14. $(3x + 13)(3x - 2) =$

15. $(c + 2d)(c - 2d) =$

16. $(3b + 6)(3b - 6) =$

17. $(2b + 5a)^2 =$

18. $(2x + 3v)^2 =$

19. $(12 + b)(12 - b) =$

20. $(5x + 7)^2 =$

© Carson-Dellosa • CD-104632

Name _____ 7.EE.A.1, HSA-SSE.A.2, HSA-APR.A.1, HSA-APR.C.4

Multiplying Binomials

Multiply.

1. $(4b^2 - 4)(4b^2 + 4) =$

2. $(3b - 3c)^2 =$

3. $(x - 2y)^2 =$

4. $(^-5x^2 + 3)(^-5x^2 - 3) =$

5. $(7b^2 - 3c)^2 =$

6. $(4c + 9d)^2 =$

7. $(5x^2 - 5y)^2 =$

8. $(x - yz)(x + yz) =$

9. $(^-4x + 3y)^2 =$

10. $(4m^2 - 2n)^2 =$

11. $(x^2 - 7y)^2 =$

12. $(4x^2 - 4y^2)^2 =$

13. $(8x^2 - 12)(8x^2 + 12) =$

14. $(2b^2 - 2c^2)^2 =$

15. $(4a + b)^2 =$

16. $(x^2 - 8x)(x^2 + 8x) =$

17. $(2x^2 - y^2)(2x^2 + y^2) =$

18. $(3x^2 - x)(3x^2 + x) =$

19. $(3a - 7b)^2 =$

20. $(^-6x + 3y)^2 =$

21. $(4x^2 - 4y^2)(4x^2 + 4y^2) =$

22. $(^-2x^3 + 4)(2x^2 + 5) =$

© Carson-Dellosa • CD-104632

Name _____ 7.EE.A.1, HSA-SSE.A.2

Factoring Polynomials

To factor a polynomial, write the polynomial as a product of other polynomials.
For example, $3x^2 - 6x$ can be written as $3x(x - 2)$.
$3x$ is the Greatest Common Factor (GCF) of $3x^2$ and $6x$.
$3x$ is a Common Monomial Factor of the terms of the binomial.
$x - 2$ is a Binomial Factor of $3x^2 - 6$.

To factor a trinomial in the form $x^2 + bx + c$, write it as the product of two binomials.
For example, $a^2 - 9a + 14$ can be written as $(a - 7)(a - 2)$.

Look for two numbers that are the product of c and whose sum is b.
If c and b are both positive, the factors will both be positive.
If c is positive and b is negative, the factors will both be negative.
If c is negative, the factors will have opposite signs.

Factor.

1. $3x^2 - 12x^3 =$

2. $2x^3 - x^4 =$

3. $3a^5 - a^3 =$

4. $x^5 + 2x^2 =$

5. $24b^2 + 16b =$

6. $5x^3 - 7x^2 =$

7. $2x^3 + 6x^2 =$

8. $x^3 - 5x^2 =$

9. $x^2 + x - 90 =$

10. $y^2 - 13y + 42 =$

11. $x^2 - x - 6 =$

12. $x^2 - 9x + 18 =$

13. $b^2 - 4b - 45 =$

14. $y^2 - 12y + 36 =$

© Carson-Dellosa • CD-104632

Name _____

7.EE.A.1, HSA-SSE.A.2

Factoring Polynomials

$$a^2 - 9a + 14 = (a - 7)(a - 2)$$
$$2x^2 - 5x - 12 = (2x + 3)(x - 4)$$

Factor.

1. $p^2 + 13p + 42 =$

2. $c^2 + c - 30 =$

3. $x^2 + 15xy + 44y^2 =$

4. $x^2 - 13x + 12 =$

5. $x^2 - 13x + 30 =$

6. $x^2 - 8x + 16 =$

7. $x^2 - 12xy + 32y^2 =$

8. $x^2 + 14x + 49 =$

9. $n^2 + 6n - 16 =$

10. $x^2 + 12x + 35 =$

11. $c^2 - 10c + 21 =$

12. $x^2 + 6x - 40 =$

13. $2x^2 + 9x + 10 =$

14. $4x^2 - 18x + 20 =$

15. $3x^2 - 10x + 7 =$

16. $3x^2 - 5x - 12 =$

17. $3x^2 - 4x - 32 =$

18. $5x^2 + 25x + 30 =$

© Carson-Dellosa • CD-104632

Name _____

7.EE.A.1, HSA-SSE.A.2

Factoring Polynomials

Factor.

1. $x^2 + 4x + 4 =$

2. $2x^2 + 13x + 6 =$

3. $x^2 + 7x + 12 =$

4. $6x^2 - 21x - 12 =$

5. $x^2 - 8x + 15 =$

6. $9x^2 - 9x - 28 =$

7. $x^2 - 15x + 56 =$

8. $15x^2 + 11x - 14 =$

9. $x^2 + 18x + 45 =$

10. $10x^2 - 28x - 6 =$

11. $x^2 + 4x - 32 =$

12. $7x^2 + 17x + 6 =$

13. $x^2 + 2xy - 63y^2 =$

14. $2x^2 - 2x - 40 =$

15. $x^2 - 14x - 72 =$

16. $11x^2 - 122x + 11 =$

17. $x^2 - xy - 2y^2 =$

18. $2x^2 + 7x + 3 =$

19. $x^2 + 16x + 28 =$

20. $12x^2 + 9x - 3 =$

© Carson-Dellosa • CD-104632

Name _____

7.EE.A.1, HSA-SSE.A.2

Factoring Polynomials—Special Cases

Factoring Trinomials That Are Quadratic in Form

$x^4 - x^2 - 12 = (x^2)^2 - (x^2) - 12 = (x^2 - 4)(x^2 + 3)$

Factoring the Difference of Two Squares

Rule: $a^2 - b^2 = (a + b)(a - b)$ Example: $x^2 - 49 = (x + 7)(x - 7)$

Factoring Perfect Square Trinomials

Rules: $a^2 + 2ab + b^2 = (a + b)^2$ $a^2 - 2ab + b^2 = (a - b)^2$

Examples: $9x^2 + 6x + 1 = (3x + 1)^2$ $x^2 - 6x + 9 = (x - 3)^2$

Factoring the Sum or Difference of Two Cubes

Rules: $x^3 + y^3 = (x + y)(x^2 - xy + y^2)$ $x^3 - y^3 = (x - y)(x^2 + xy + y^2)$

Examples: $x^3 + 8 = (x + 2)(x^2 - 2x + 4)$ $x^3 - 8 = (x - 2)(x^2 + 2x + 4)$

Factor.

1. $2x^2 - 5x - 12 =$

2. $2x^4 + 16x^2 + 30 =$

3. $x^4 - 8x^2 + 15 =$

4. $7x^4 - 11x^2 - 6 =$

5. $a^2 - 4 =$

6. $b^2 - 9 =$

7. $1 - 9x^2 =$

8. $x^2 - 25 =$

9. $x^2 + 14x + 49 =$

10. $x^2 - 2x + 1 =$

11. $c^2 - 6c + 9 =$

12. $x^2 - 4xy + 4y^2 =$

13. $64x^3 + 1 =$

14. $8x^3 + 27 =$

© Carson-Dellosa • CD-104632

Name _____

7.EE.A.1, HSA-SSE.A.2

Factoring Polynomials—Special Cases

Factor.

1. $2x^4 - 7x^2 - 15 =$

2. $y^4 + 6y^2 - 16 =$

3. $8x^4 - 23x^2 - 3 =$

4. $6a^6 - 5a^3b^3 - 25b^6 =$

5. $3x^4 + 20x^2 + 33 =$

6. $4x^4y^4 - 2x^2y^2 - 56 =$

7. $y^2 - 81 =$

8. $c^2 - 16 =$

9. $a^2 - 49 =$

10. $49x^2 - 16y^4 =$

11. $16x^2 - 121 =$

12. $25 - x^2y^2 =$

13. $a^2 + 12ab + 36b^2 =$

14. $49x^2 + 28x + 4 =$

15. $x^2 + 14x + 49 =$

16. $c^2 - 20c + 100 =$

17. $x^3 - 1000 =$

18. $1 - 64y^3 =$

19. $x^3 + 125 =$

20. $x^3 - 27 =$

© Carson-Dellosa • CD-104632

Name _____

7.EE.A.1, HSA-SSE.A.2

Factoring Polynomials—Special Cases

Factor.

1. $y^4 - y^2 - 12 =$

2. $x^4y^4 - 19x^2y^2 + 34 =$

3. $2x^4y^4 - 17x^2y^2 - 30 =$

4. $x^4y^4 - 8x^2y^2 + 12 =$

5. $64 - x^4y^4 =$

6. $y^2 - 64 =$

7. $81x^2 - 4 =$

8. $16 - 81x^2 =$

9. $x^2y^2 - 121 =$

10. $49x^2 - 36 =$

11. $y^2 - 22y + 121 =$

12. $25a^2 - 40ab + 16b^2 =$

13. $9x^2 - 12x + 4 =$

14. $16x^2 - 40x + 25 =$

15. $9x^2 + 12x + 4 =$

16. $x^2 - 14x + 49 =$

17. $27x^3 - 64 =$

18. $27x^3 - 27 =$

19. $125a^3 - 8b^3 =$

20. $64x^3 - y^3 =$

21. $125x^3 - 64y^3 =$

22. $64x^3 + 27 =$

© Carson-Dellosa • CD-104632

Name _____ HSA-SSE.B.3a, HSA-REI.B.4b

Solving Equations by Factoring

The **Multiplication Property of Zero**: The product of a number and zero is zero.

The **Principle of Zero Products**: If the product of two factors is zero, then at least one of the factors must be zero. This principle is used in solving equations.

Solve: $(x - 4)(x - 5) = 0$

$x - 4 = 0 \quad x - 5 = 0$

$x = 4 \quad x = 5$

The solutions are 4 and 5.

If $(x - 4)(x - 5) = 0$, then $(x - 4) = 0$ or $(x - 5) = 0$.

$x = 4$	$x = 5$
$(4 - 4)(4 - 5) = 0$	$(5 - 4)(5 - 5) = 0$
$(0)(^{-}1) = 0$	$(1)(0) = 0$
$0 = 0$	$0 = 0$

Write the solutions for each variable.

1. $x(x + 6) = 0$
2. $b^2 - 81 = 0$
3. $(27 - y)(y - 2) = 0$
4. $z^2 - 1 = 0$
5. $(y - 4)(y - 8) = 0$
6. $y(y - 11) = 0$
7. $8b^2 - 32 = 0$
8. $x^2 - x - 6 = 0$
9. $x^2 - 4x - 21 = 0$
10. $m^2 - 144 = 0$
11. $(y + 5)(y + 6) = 0$
12. $(2x + 4)(x + 7) = 0$
13. $z^2 - 9 = 0$
14. $10x^2 - 10x = 0$
15. $2x^2 - 6x = x - 3$
16. $4y(3y - 2) = 0$
17. $(4y - 1)(y + 2) = 0$
18. $x^2 - 5x + 6 = 0$

© Carson-Dellosa • CD-104632

Name _____

HSA-SSE.B.3a, HSA-REI.B.4b

Solving Equations by Factoring

$x^2 - 8x = {}^-16$	$x - 4 = 0$
$x^2 - 8x + 16 = 0$	$x = 4$
$(x - 4)(x - 4) = 0$	The solution is 4.

Solve by factoring.

1. $y^2 + 9y = 0$

2. $x - 16 = x(x - 7)$

3. $x - 6 = x(x - 4)$

4. $x^2 + 7x = 0$

5. $x^2 - 4x = 0$

6. $x + 8 = x(x + 3)$

7. $y^2 - y - 6 = 0$

8. $a^2 - 36 = 0$

9. $y^2 + 15 = 8y$

10. $a^2 - 7a = {}^-12$

11. $y^2 + 36y = 0$

12. $3u^2 - 12u - 15 = 0$

13. $y^2 - 8y + 12 = 0$

14. $5a^2 + 25a = 0$

15. $6x^2 + 18x = 0$

16. $2x^2 + x = 6$

17. $x^2 - 5x - 6 = 0$

18. $4x^2 + 16x = 0$

19. $3x^2 - 9x = 0$

20. $y^2 + 5y - 6 = 0$

© Carson-Dellosa • CD-104632

Name _____

HSA-SSE.B.3a, HSA-REI.B.4b

Solving Equations by Factoring

Solve by factoring.

1. $a^2 - 8a = 0$

2. $x^2 = 3x + 4$

3. $4a^2 + 15a - 4 = 0$

4. $x^2 - x - 6 = 0$

5. $3x^2 - 13x + 4 = 0$

6. $6x^2 = 23x + 18$

7. $x^2 + 7x + 12 = 0$

8. $x^2 + 5x - 6 = 0$

9. $x^2 = 6x + 7$

10. $x^2 = 10x - 25$

11. $x^2 + 3x - 10 = 0$

12. $x^2 - 6x + 9 = 0$

13. $y^2 - 3y + 2 = 0$

14. $2x^2 - 9x + 9 = 0$

15. $r^2 - 15r = 16$

16. $x^2 + 7x + 10 = 0$

17. $3x^2 - 2x - 8 = 0$

18. $2a^2 + 4a - 6 = 0$

19. $x^2 + 3x - 4 = 0$

20. $4a^2 + 9a + 2 = 0$

21. $9x^2 = 18x + 0$

22. $2x^2 = 9x + 5$

© Carson-Dellosa • CD-104632

Name _____ 7.EE.B.4, HSA-SSE.B.3a, HSA-REI.B.4b

Writing and Solving Quadratic Equations

Problem solving often involves quadratic equations.

The square of a number is 8 less than 6 times the number. Find the numbers that make the sentence true.

$x^2 = 6x - 8$	First, set up the equation.
$x^2 - 6x + 8 = 0$	Put the equation in standard form.
$(x - 2)(x - 4) = 0$	Factor.
$x - 2 = 0$ or $x - 4 = 0$	Solve for x.
$x = 2$ or $x = 4$	So, the numbers are 2 and 4.

The product of a number and 10 more than 4 times the number is 50. Find the numbers that make the sentence true.

$x(4x + 10) = 50$	First, set up the equation.
$4x^2 + 10x - 50 = 0$	Put the equation in standard form.
$2(2x - 5)(x + 5) = 0$	Factor.
Since $2 \neq 0$, then $2x - 5 = 0$ or $x + 5 = 0$	Solve for x.
$x = \frac{5}{2}$ and $x = {}^{-}5$	So, the numbers are $\frac{5}{2}$ and ${}^{-}5$.

Write an equation for each word problem and solve it.

1. Six less than 3 times a number is 12 less than twice the number.

   Equation _____ **Solution** _____

2. The square of a number is 8 more than twice the number.

   Equation _____ **Solution** _____

3. The sum of the square of a number and 4 times the number is 12.

   Equation _____ **Solution** _____

4. Twice the square of a number is 4 more than twice the number.

   Equation _____ **Solution** _____

5. Three times the square of a number is 5 less than 16 times the number.

   Equation _____ **Solution** _____

© Carson-Dellosa • CD-104632

Name _____ 7.EE.B.4, HSA-SSE.B.3a, HSA-REI.B.4b

Writing and Solving Quadratic Equations

The length of a rectangle is 3 inches longer than the width. The area of the rectangle is 40 square inches. Find the length and width of the rectangle.

Width of rectangle: w
Length of rectangle: $w + 3$
$A = lw$
$40 = (w + 3)(w)$
$40 = w^2 + 3w$
$0 = w^2 + 3w - 40$
$0 = (w + 8)(w - 5)$
$w + 8 = 0$ or $w - 5 = 0$
$w = {}^{-}8 \quad w = 5$

Since the width cannot be a negative number, the width is 5.

$l = 5 + 3 = 8$

The length is 8 inches, and the width is 5 inches.

Write an equation for each word problem and solve it.

1. The area of a square is 121 m^2. Find the length of the sides of the square.

   Equation _____ **Solution** _____

2. The area of a rectangle is 72 m^2. Its length is twice its width. Find the length and width of the rectangle.

   Equation _____ **Solution** _____

3. The area of a rectangle is 36 cm^2. Its width is 4 times its length. Find the length and width of the rectangle.

   Equation _____ **Solution** _____

4. The width of a rectangle is 5 more than twice its length. The area of the rectangle is 33 $in.^2$. Find the dimensions of the rectangle.

   Equation _____ **Solution** _____

5. The length of a rectangle is 4 more than twice its width. The area of the rectangle is 96 $ft.^2$. Find its dimensions.

   Equation _____ **Solution** _____

© Carson-Dellosa • CD-104632

Name _____ 7.EE.B.4, HSA-SSE.B.3a, HSA-REI.B.4b

Writing and Solving Quadratic Equations

Write an equation for each word problem and solve it.

1. The length of a rectangle is 5 inches longer than the width. The area of the rectangle is 50 square inches. Find the length and width.

   Equation _____ **Solution** _____

2. The length of a rectangle is 2 more than 8 times its width. Its area is 36 cm^2. Find the width and length.

   Equation _____ **Solution** _____

3. The sum of a number and its square is 42. Find the numbers.

   Equation _____ **Solution** _____

4. The sum of a number and its square is 56. Find the numbers.

   Equation _____ **Solution** _____

5. The length of a rectangle is twice its width. Its area is 32 m^2. Find the length and width.

   Equation _____ **Solution** _____

6. The square of a number is 80 more than 2 times the number. Find the numbers.

   Equation _____ **Solution** _____

7. The square of a number is 48 more than 2 times the number. Find the numbers.

   Equation _____ **Solution** _____

8. The width of a rectangle is 2/3 the length. The area is 54 in^2. Find the length and width.

   Equation _____ **Solution** _____

© Carson-Dellosa • CD-104632

Name _____

HSA-APR.D.6

Dividing Polynomials

Simplify using long division: $(x^2 + 6x + 5) \div (x + 1)$

$$\frac{(x^2 + 6x + 5)}{(x + 1)}$$

$$\begin{array}{r} x + 5 \\ x + 1 \overline{)\, x^2 + 6x + 5} \\ \underline{-\, x^2 + 1x } \\ 5x + 5 \\ \underline{-\, 5x + 5} \\ 0 \end{array}$$

Divide by using long division.

1. $(x^2 + 5x + 6) \div (x + 3) =$

2. $(x^2 + 4x - 21) \div (x - 3) =$

3. $(x^2 - 3x - 40) \div (x + 5) =$

4. $(x^2 - x - 42) \div (x + 6) =$

5. $(x^2 - 8x + 16) \div (x - 4) =$

6. $(x^2 + 2x - 35) \div (x + 7) =$

7. $(x^2 - 6x + 9) \div (x - 3) =$

8. $(x^2 + 5x + 4) \div (x + 1) =$

9. $(x^2 + 7x + 10) \div (x + 2) =$

10. $(x^2 + 9x + 8) \div (x + 8) =$

© Carson-Dellosa • CD-104632

Name _____ HSA-APR.D.6

Dividing Polynomials

Simplify using synthetic division: $(2x^3 + 3x^2 - 4x + 8) \div (x + 3)$

$$\begin{array}{r|rrrr} -3 & 2 & 3 & -4 & 8 \\ & & -6 & 9 & -15 \\ \hline & 2 & -3 & 5 & -7 \end{array}$$

$$= 2x^2 - 3x + 5 - \frac{7}{x+3}$$

Divide by using synthetic division.

1. $(x^2 + x - 2) \div (x + 2) =$

2. $(5x^2 - 12x - 9) \div (x - 3) =$

3. $(3x^2 - 5) \div (x - 1) =$

4. $(3x^3 + 8x^2 + 9x + 10) \div (x + 2) =$

5. $(x^3 - 4x^2 - 36x - 16) \div (x + 4) =$

6. $(3x^2 - 7x + 6) \div (x - 3) =$

7. $(4x^2 + 9x + 6) \div (x + 1) =$

8. $(4x^2 + 23x + 28) \div (x + 4) =$

9. $(3x^2 + 19x + 20) \div (x + 5) =$

10. $(x^2 + 14x + 45) \div (x + 5) =$

© Carson-Dellosa • CD-104632

Name _____

HSA-APR.D.6

Dividing Polynomials

Divide.

1. $\frac{x^2 + 9x + 8}{x - 1} =$

2. $\frac{2x^2 - 5x - 3}{x - 3} =$

3. $\frac{4x^2 - 7x - 2}{4x + 1} =$

4. $\frac{2x^2 + x - 3}{x - 1} =$

5. $\frac{21x^2 + 22x - 8}{3x + 4} =$

6. $\frac{3x^2 - 27}{x + 3} =$

7. $\frac{9x^2 - 27x - 36}{x - 4} =$

8. $\frac{5x^2 + 43x - 18}{x + 9} =$

9. $\frac{x^2 - 4x - 45}{x + 5} =$

10. $\frac{2x^2 - x - 21}{x + 3} =$

11. $(3x^3 - 13x^2 - 13x + 15) \div (x - 5) =$

12. $(2x^3 - 12x^2 + 5x - 27) \div (x - 6) =$

13. $(3x^2 - 75) \div (x - 5) =$

14. $(2x^2 + 7x - 10) \div (x + 1) =$

© Carson-Dellosa • CD-104632

Name _____

HSA-APR.D.7

Operations with Rational Expressions

Add or subtract:

$$\frac{7x-12}{2x^2+5x-12}-\frac{3x-6}{2x^2+5x-12}=\frac{(7x-12)-(3x-6)}{2x^2+5x-12}=\frac{4x-6}{(2x-3)(x+4)}=$$

$$\frac{2\cancel{(2x-3)}}{\cancel{(2x-3)}(x+4)}=\frac{2}{x+4}$$

Multiply:

$$\frac{4x+16}{12x+48}\bullet\frac{(x-4)}{2x}=\frac{\cancel{4(x+4)}(x-4)}{\cancel{24}x\cancel{(x+4)}}=\frac{x-4}{6x}$$

Divide:

$$\frac{x+2}{x+3}\div\frac{x^2-4}{x-2}=\frac{\cancel{x+2}}{x+3}\bullet\frac{\cancel{x-2}}{\cancel{(x+2)}\cancel{(x-2)}}=\frac{1}{x+3}$$

Simplify.

1. $\frac{x^2+x-6}{x+1}\bullet\frac{x+1}{x^2-9}=$

2. $\frac{x^2-1}{x^2-2x-3}\bullet\frac{x+4}{6x-6}=$

3. $\frac{x-7}{x+2}\div\frac{x^2-49}{x^2+9x+14}=$

4. $\frac{x+2}{4x(x-6)}\div\frac{x^2-4}{8x(x-6)}=$

5. $\frac{2}{x+2}+\frac{6x}{x^2-4}=$

6. $\frac{2}{4xy}+\frac{14}{3xy}-\frac{9}{2xy}=$

© Carson-Dellosa • CD-104632

Name _____

HSA-APR.D.7

Operations with Rational Expressions

Simplify.

1. $\frac{x+2}{x-3} \bullet \frac{x^2-8x+15}{5x-25} =$

2. $\frac{12x^2y^4}{36ab^3} \bullet \frac{6a^2v^3}{48xy^4} =$

3. $\frac{x^2+6x+8}{x^2-16} \bullet \frac{3x-12}{4x+4} =$

4. $\frac{x^2-2x-8}{x^2-4} \bullet \frac{x-2}{x+3} =$

5. $\frac{x^2+3x+2}{x+7} \bullet \frac{x^2+9x+14}{x^2+4x+4} =$

6. $\frac{2x^2+12x+18}{x^2+5x-6} \div \frac{2x+6}{x-1} =$

7. $\frac{2x^2+6x}{x^2+2x} \div \frac{x^2-9}{4x-12} =$

8. $\frac{15x^4y^2}{5xy} \div \frac{10x^3y}{5y^2} =$

9. $\frac{x^2+8x}{x^2+14x+48} \div \frac{x^2+x}{x^2} =$

10. $\frac{3x^2+6x}{x^2+6x} \div \frac{x^2-4}{2x-4} =$

11. $\frac{x}{x-4} + \frac{4}{x^2-x-12} =$

12. $\frac{4}{3x-8} - \frac{x}{4x-7} =$

13. $-\frac{4}{2x^2} + \frac{5}{2x^2} + \frac{8}{3x^2} =$

14. $\frac{4x}{x^2+x-2} - \frac{4x}{x^2+x-2} =$

© Carson-Dellosa • CD-104632

Name _____ HSA-APR.D.7

Operations with Rational Expressions

Simplify.

1. $\frac{2x^2 - 32}{2x + 8} \bullet \frac{x^2 - 9}{x^2 - 3x - 4} =$

2. $\frac{14x^2y^4}{42a^2b^4} \bullet \frac{28a^2b^3}{35x^3y^4} =$

3. $\frac{x^2 - 100}{x - 5} \bullet \frac{x + 5}{x^2 - 5x - 50} =$

4. $\frac{x^2 - 12x + 35}{x^2 - 5x - 14} \bullet \frac{x^2 + 7x + 10}{x^2 - 25} =$

5. $\frac{9x^2 - 25}{4x^2 + 4x - 3} \bullet \frac{40x^2 - 10x}{3x + 5} =$

6. $\frac{x + 4}{6x^2 + 24} \bullet \frac{2x^3 - 8x}{x^2 + 4x} =$

7. $\frac{x^2 - 7x}{x^2 - 14x + 49} \div \frac{2x^2 + 6x}{x^2 + x - 56} =$

8. $\frac{x^2 - 16}{x^2 + 7x + 12} \div \frac{5x - 20}{x + 3} =$

9. $\frac{27x^6y^2}{9x^3y} \div \frac{4x^5y^3}{16x^4y^2} =$

10. $\frac{x^2 + 9x - 10}{x^2 + 5x - 14} \div \frac{3x + 30}{2x - 4} =$

11. $\frac{24a^4b^2}{8a^2b} \div \frac{12ab^3}{16a^3b} =$

12. $\frac{11x}{x^2 - 6x - 7} + \frac{5x}{x^2 + 9x + 8} =$

13. $\frac{3x}{x^2 - 4} + \frac{5x}{x^2 - 4} =$

14. $-\frac{10}{4x^2} + \frac{6}{4x^2} + \frac{8}{4x^2} =$

15. $\frac{5}{2xy} + \frac{5}{4xy} - \frac{12}{6xy} =$

16. $\frac{8}{4x + 16} - \frac{1}{x + 4} - \frac{3}{4} =$

© Carson-Dellosa • CD-104632

Name _____

7.RP.A.1

Ratios and Proportions

Solve the following ratio for x.

$\frac{x}{5} = \frac{4}{10}$ → Take cross products and solve. → $\frac{x}{5} = \frac{4}{10}$ $5 \bullet 4 = 20$, $x \bullet 10 = 10x$

→ $10x = 20$ → $\frac{10x}{10} = \frac{20}{10}$ → $x = 2$

Solve.

1. $\frac{4}{(x-3)} = \frac{28}{49}$

2. $\frac{(5+x)}{10} = \frac{2}{5}$

3. $\frac{x}{30} = \frac{7}{10}$

4. $\frac{(x-2)}{16} = \frac{x}{4}$

5. $\frac{2}{x} = \frac{6}{30}$

6. $\frac{(x+1)}{7} = \frac{6}{14}$

7. $\frac{x}{15} = \frac{5}{75}$

8. $\frac{x}{20} = \frac{2}{10}$

9. $\frac{x}{6} = \frac{(x-3)}{12}$

10. $\frac{x}{5} = \frac{12}{6}$

11. $\frac{6}{(x+5)} = \frac{18}{24}$

12. $\frac{5}{15} = \frac{x}{9}$

13. $\frac{+10}{} = \frac{5}{2}$

14. $\frac{x}{3} = \frac{12}{27}$

© Carson-Dellosa • CD-104632

Name _____

6.RP.A.3, 7.RP.A.1, 7.RP.A.2c, 7.RP.A.3

Ratios and Proportions

Three liters of soda cost $3.00. At this rate, how much would 10 liters of soda cost? To find the cost, write and solve a ratio using x to represent the cost.

$\frac{\text{liters}}{\text{cost}} \longrightarrow \frac{3}{3.00} = \frac{10}{x} \longrightarrow 3x = 10(3.00) \longrightarrow 3x = 30 \longrightarrow \frac{3x}{3} = \frac{30}{3}$

$\longrightarrow x = 10 \longrightarrow$ The cost of 10 liters of soda is $10.00.

Solve.

1. $\frac{4}{x+8} = \frac{3}{x}$

2. $\frac{x}{4} = \frac{21}{7}$

3. $\frac{9}{3} = \frac{6}{x}$

4. $\frac{x-3}{4} = \frac{x-4}{3}$

5. $\frac{x}{4} = \frac{8}{2x}$

6. $\frac{12}{x} = 3$

Set up a proportion and solve.

7. A copy machine can print 120 pages per minute. At this rate, how many minutes will it take to make 840 copies?

8. Two gallons of fruit juice will serve 35 people. How much fruit juice is needed to serve 105 people?

9. A stock investment of $4,000 earns $360 in interest each year. At the same rate, how much interest will a person earn if he invests $6,000?

© Carson-Dellosa • CD-104632

Name _____

6.RP.A.3, 7.RP.A.1, 7.RP.A.2c, 7.RP.A.3

Ratios and Proportions

Set up a proportion and solve.

1. An investment of $36,000 earns $900 each year. At the same rate, how much money must be invested to earn $1,200 each year?

2. The sales tax on a $15,000 car is $540. At this rate, what is the tax on a $32,000 car?

3. Six gallons of paint will cover 120 doors. At this rate, how many gallons of paint are needed to cover 480 doors?

4. A lawnmower can cut 1 acre on 0.5 gallons of gasoline. At this rate, how much gasoline is needed to cut 3.5 acres?

5. An aerobics instructor burns 400 calories in 1 hour. How many hours would the instructor have to do aerobics to burn 660 calories?

6. The real estate tax for a house that costs $56,000 is $1,400. At this rate, what is the value of a house for which the real estate tax is $1,800?

7. One hundred thirty-six tiles are required to tile a 36 $ft.^2$ area. At this rate, how many tiles are required to tile a 288 $ft.^2$ area?

© Carson-Dellosa • CD-104632

Name _____

5.G.A.1, 6.NS.C.6b, 6.NS.C.6c

Graphing Ordered Pairs

Start at the origin (0, 0).

(x, y) = (1, ⁻2) right 1 and down 2

(x, y) = (⁻3, 4) left 3 and up 4

(x, y) = (2, ⁻2) right 2 and down 2

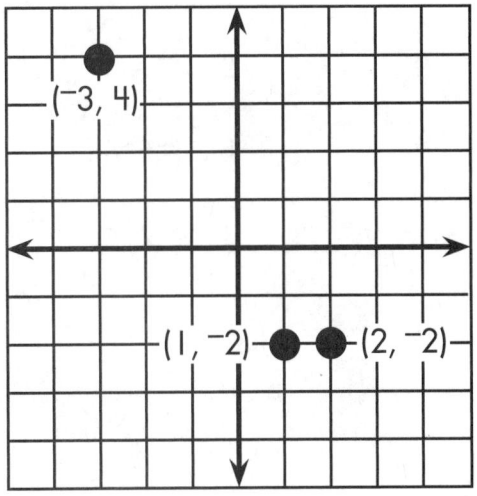

Label the following points. Start at the origin (0, 0).

A (6, 2)

B (6, 3)

C (5, 1)

D (5, 0)

E (1, ⁻3)

F (2, ⁻6)

G (5, 4)

H (2, 4)

I (4, 3)

J (1, ⁻2)

K (5, ⁻3)

L (1, 4)

M (3, 1)

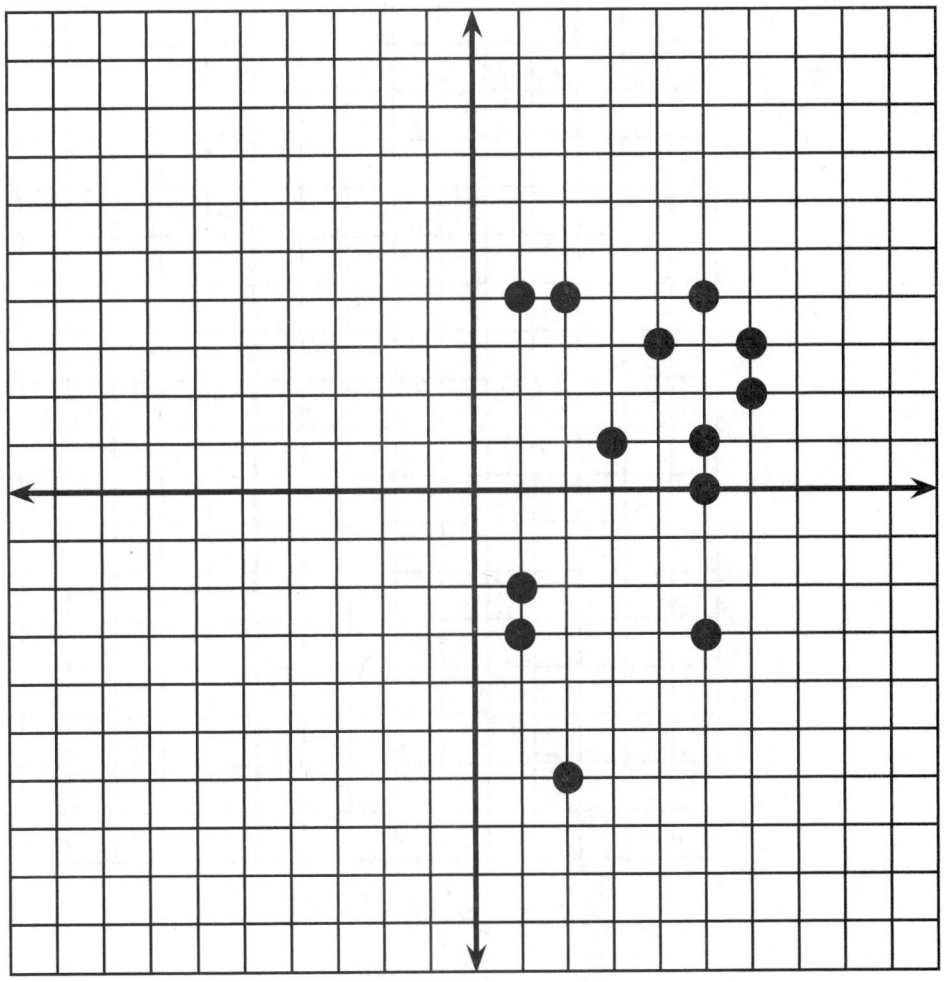

© Carson-Dellosa • CD-104632

Name _____

5.G.A.1, 6.NS.C.6b, 6.NS.C.6c

Graphing Ordered Pairs

Graph and label the following points.

A $(^{-}1, 2)$ B $(^{-}1, 0)$ C $(^{-}1, ^{-}6)$ D $(^{-}4, 2)$

E $(^{-}6, 2)$ F $(^{-}5, 6)$ G $(^{-}4, 1)$ H $(2, ^{-}8)$

I $(6, 6)$ J $(^{-}7, 5)$ K $(^{-}1, ^{-}1)$ L $(^{-}3, ^{-}3)$

M $(6, ^{-}4)$

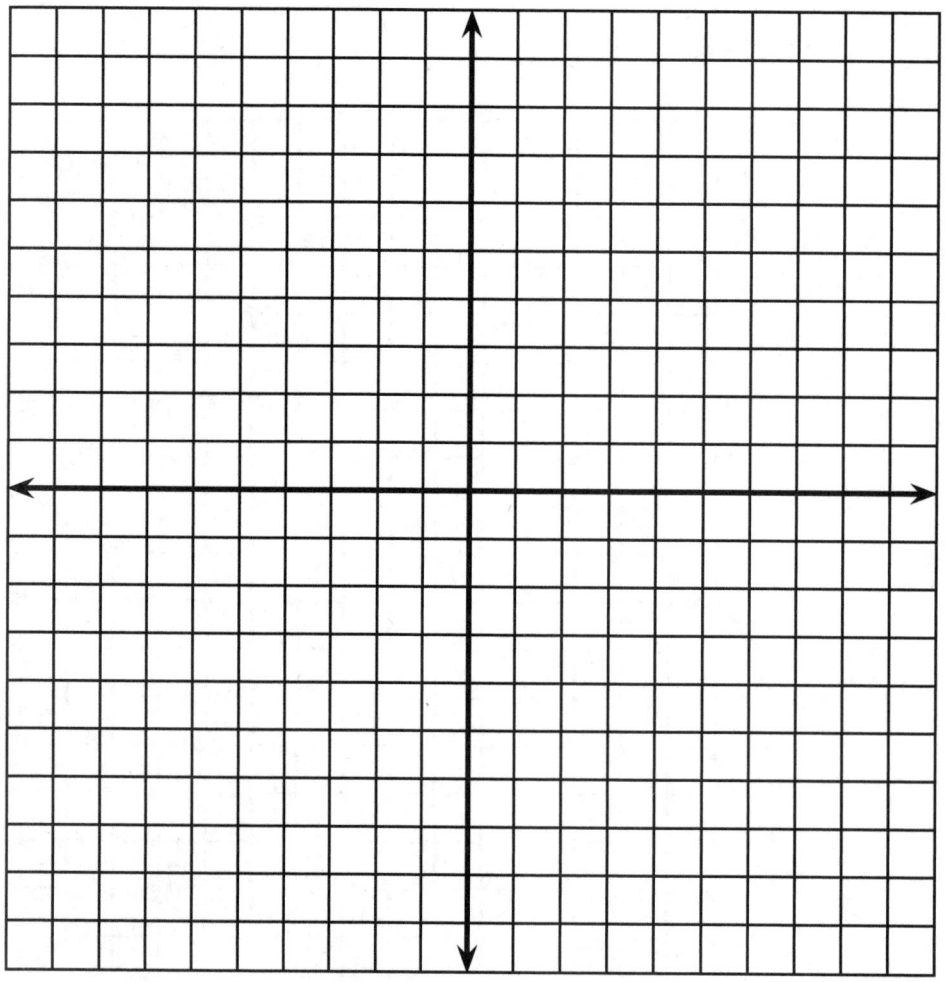

© Carson-Dellosa • CD-104632

Name _____

6.NS.C.6, 6.EE.B.6, 8.F.A.1, HSA-REI.D.10

Graphing Ordered Pairs

Solve for each x value. Graph each ordered pair.

1. $x + y = 4$

x	y
4	
5	
6	

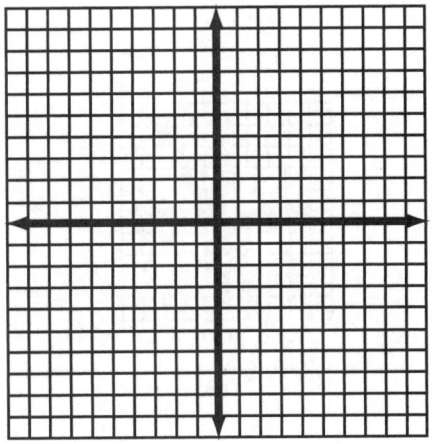

2. $y = 3 - x$

x	y
5	
−2	
0	

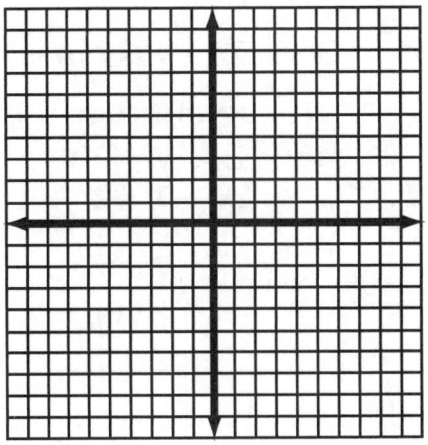

3. $4x + y = 6$

x	y
0	
1	
2	

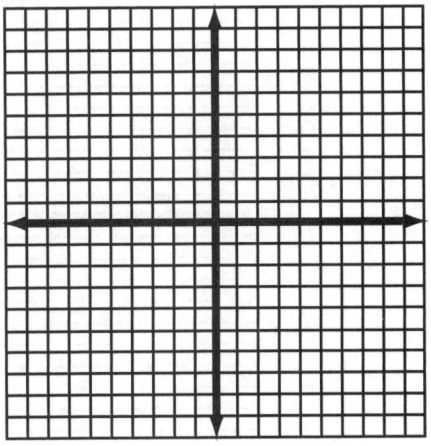

4. $y = 2x - 7$

x	y
3	
4	
5	

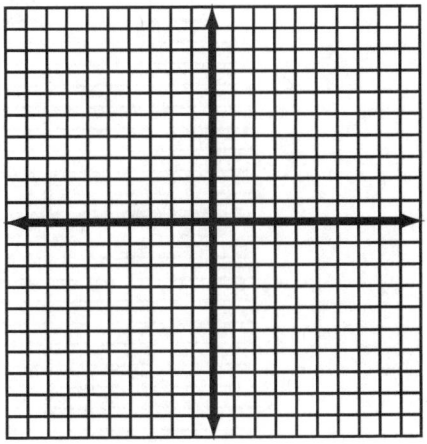

5. $y = x + 3$

x	y
2	
0	
−3	

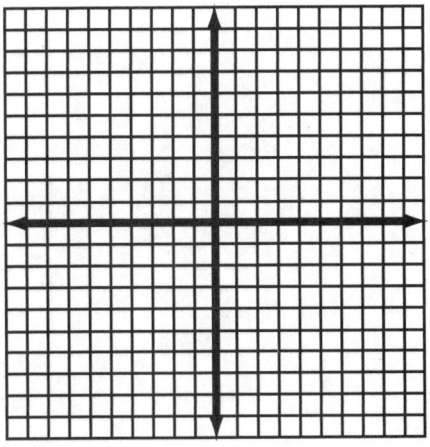

6. $y = 9 - x$

x	y
0	
4	
3	

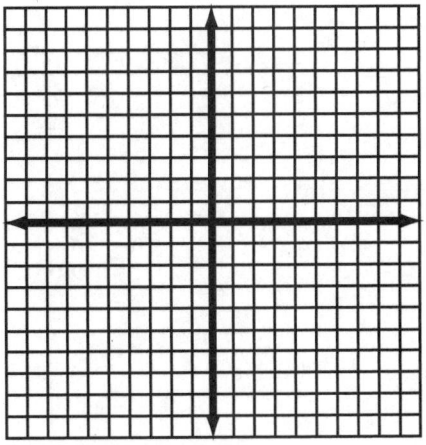

© Carson-Dellosa • CD-104632

Name _____

8.EE.B.6, 8.F.A.1

Graphing Linear Equations

One way to graph a linear equation is to find the x- and y-intercepts. To find the x-intercept, let $y = 0$. To find the y-intercept, let $x = 0$. Plot the two points and connect with a straight line.

Graph $2x - y = 2$ by using the x- and y-intercepts.

x-intercept	**y-intercept**
$2x - y = 2$	$2x - y = 2$
$2x - 0 = 2$	$2(0) - y = 2$
$2x = 2$	$^{-}y = 2$
$x = 1$	$y = {}^{-}2$
$(1, 0)$	$(0, {}^{-}2)$

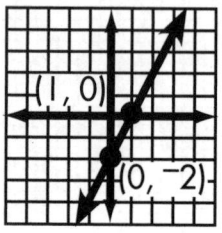

Find the x- and y-intercepts and graph.

1. $3x - 9y = 18$

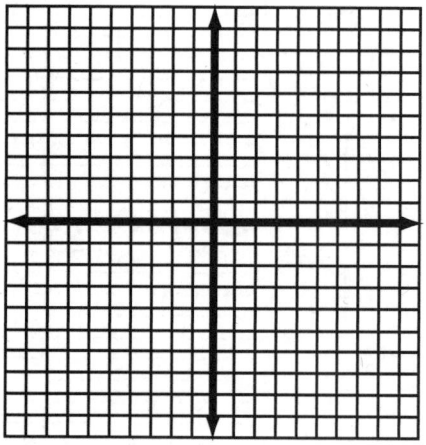

2. $4x + 2y = {}^{-}8$

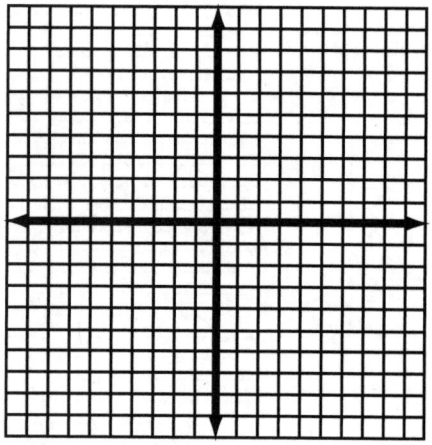

3. $2x + 4y = 20$

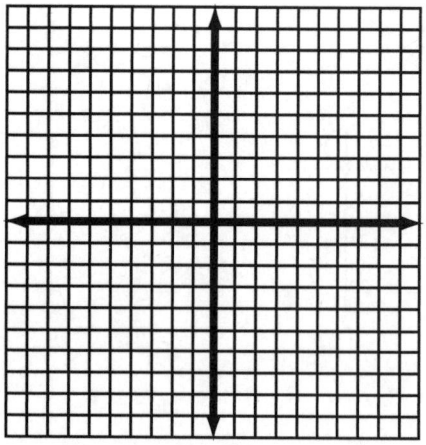

4. $x + 5y = 10$

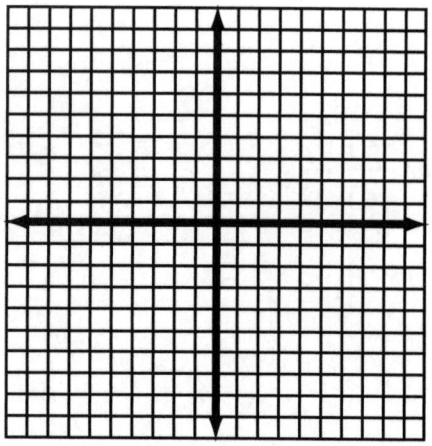

© Carson-Dellosa • CD-104632

Name _____ 8.EE.B.6, 8.F.A.1

Graphing Linear Equations

Another way to graph a linear equation is to use the **slope-intercept form**. A linear equation written in slope-intercept form is $y = mx + b$, where m is the slope of the line and b is the y-intercept.

Example: $y = 3x + 2$ $\quad$ $y = {}^{-}5x + 1$ $\quad$ $y = x - 7$

$m = 3$, y-int. $= 2$ $\quad$ $m = {}^{-}5$, y-int. $= 1$ $\quad$ $m = 1$, y-int. $= {}^{-}7$

You can rewrite an equation in slope-intercept form by adding or subtracting terms from both sides and simplifying.

To graph the equation, plot the y-intercept. Starting at the y-intercept, move up or down and then right or left according to the slope and make a new point.

Example: If $m = 4$ and $b = 3$, the y-intercept is (0, 3). The next point to graph would be (1, 7) since the slope is 4, or $\frac{4}{1}$.

Solve for y, state the m and b, and graph.

1. $y = {}^{-}3x + 5$

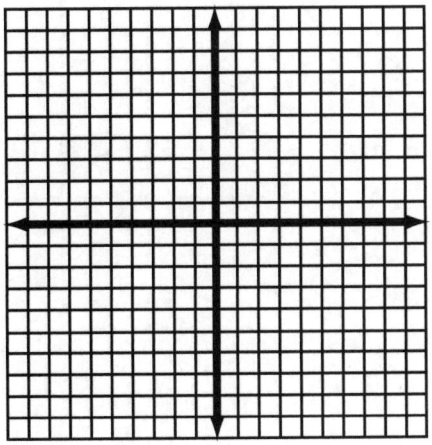

2. ${}^{-}8x + y = {}^{-}4$

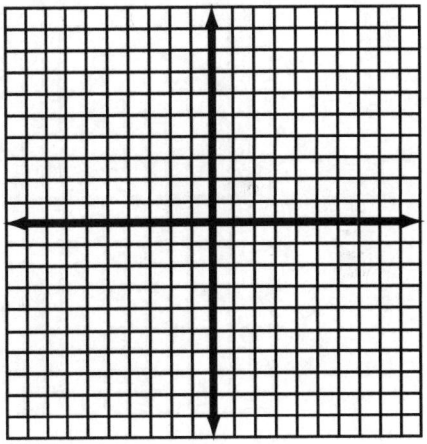

3. $x - y = 7$

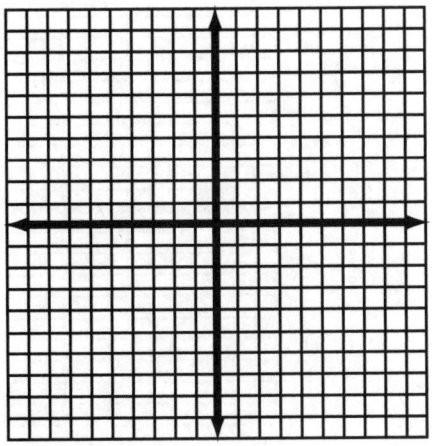

4. $2y = 3x - 4$

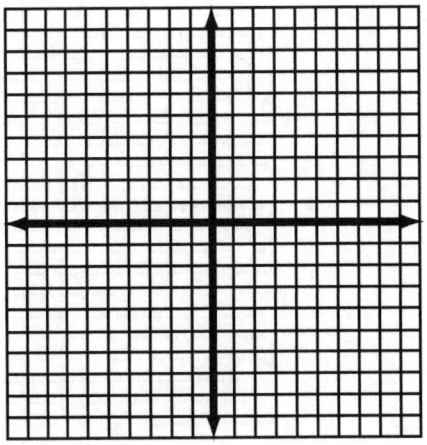

© Carson-Dellosa • CD-104632

Name _____

8.EE.B.6, 8.F.A.1

Graphing Linear Equations

Graph each equation using either slope-intercept form or x- and y-intercepts.

1. $2x + 3y = 9$

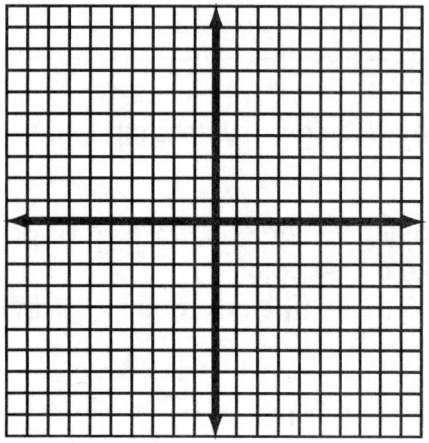

2. $x - 3y + 9 = 0$

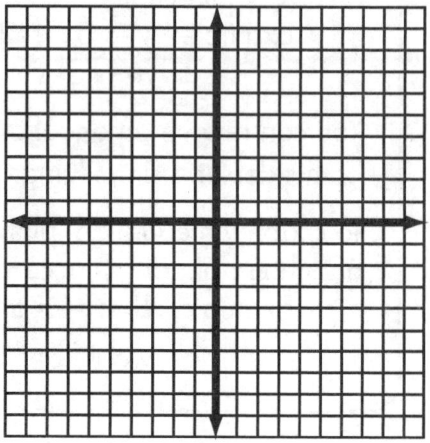

3. $6x + 2y = 12$

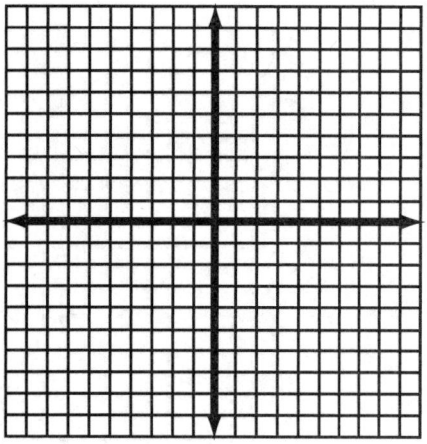

4. $y = 2x - 4$

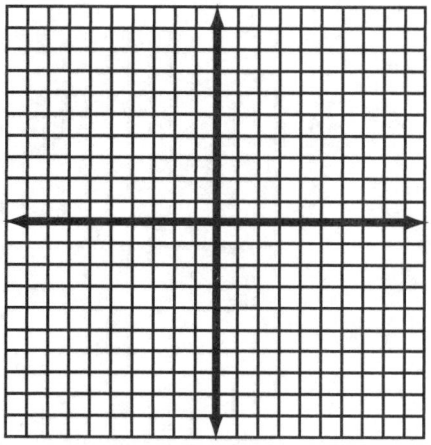

5. $5x - y = 7$

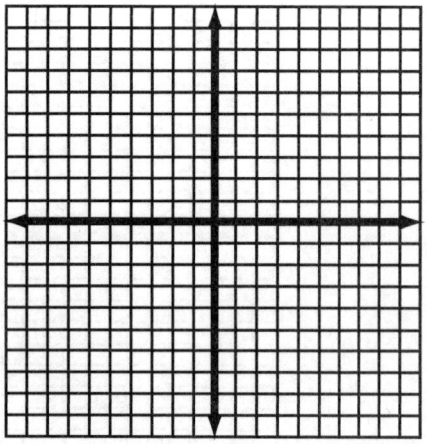

6. $x - 3y + 6 = 0$

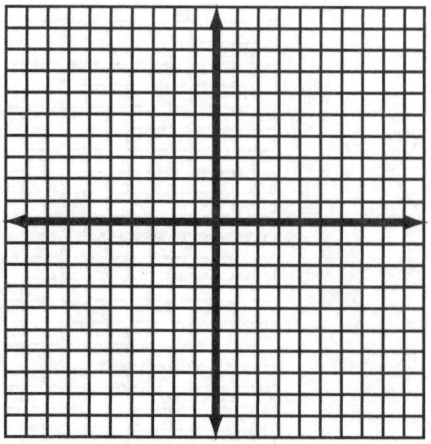

© Carson-Dellosa • CD-104632

Name _____

8.EE.B.6, 8.F.A.3

Writing Linear Equations

This equation, $y = mx + b$, is the **slope-intercept of a straight line.**
For all equations of the form $y = mx + b$, m is the slope of the line.
The y-intercept is $(0, b)$.

Find the equation of a line using the slope-intercept form: $y = mx + b$.

1. $m = 4$, y-int. $= {}^{-}1$

2. $m = {}^{-}2$, y-int. $= {}^{-}5$

3. $m = 1$, y-int. $= 2$

4. $m = \frac{7}{5}$ $b = {}^{-}2$

5. $m = \frac{3}{4}$ $b = \frac{2}{3}$

6. $m = \frac{3}{7}$ $b = \frac{1}{3}$

7. $m = \frac{3}{5}$ $b = \frac{1}{5}$

8. $m = {}^{-}4$ $b = \frac{3}{4}$

9. $m = 0$ $b = 5$

10. $m = \frac{1}{4}$ $b = \frac{2}{3}$

© Carson-Dellosa • CD-104632

Name _____

8.EE.B.6, 8.F.A.3

Writing Linear Equations

Slope-Intercept Formula
$y = mx + b$
m: slope
b: y-intercept containing coordinate points (o, b)

$m = 4$, passing through points $(1, 2)$
Using this information, $\{m = 4, x = 1, y = 2\}$.
Substitute into $y = mx + b$ to find b.

$$2 = (4)(1) + b \longrightarrow 2 = 4 + b$$
$$\longrightarrow 2 - 4 = 4 - 4 + b \longrightarrow {}^{-}2 = b$$

With the information $m = 4$, $b - {}^{-}2$, write the equation as $y = 4x - 2$.

Find the equation of the line with the given slope passing through the indicated point (P).

1. $p = ({}^{-}2, {}^{-}6)$, $m = 3$
   $b =$
   Equation _____

2. $p = ({}^{-}4, 3)$, $m = 1$
   $b =$
   Equation _____

3. $p = (3, 5)$, $m = 0$
   $b =$
   Equation _____

4. $p = (5, 7)$, $m = 3$
   $b =$
   Equation _____

5. $p = ({}^{-}7, {}^{-}7)$, $m = {}^{-}7$
   $b =$
   Equation _____

6. $p = (2, {}^{-}6)$, $m = 4$
   $b =$
   Equation _____

7. $p = (2, 4)$, $m = 4$
   $b =$
   Equation _____

8. $p = (6, {}^{-}1)$, $m = {}^{-}5$
   $b =$
   Equation _____

9. $p = ({}^{-}1, 1)$, $m = 2$
   $b =$
   Equation _____

10. $p = ({}^{-}1, {}^{-}6)$, $m = 2$
    $b =$
    Equation _____

11. $p = (4, 5)$, $m = {}^{-}2$
    $b =$
    Equation _____

12. $p = (2, 6)$, $m = {}^{-}6$
    $b =$
    Equation _____

© Carson-Dellosa • CD-104632

Name _____

8.EE.B.6, 8.F.A.3

Writing Linear Equations

Write the equation of the line that passes through the point and has the given slope. Put each equation in slope-intercept form.

1. $p = (2, 3), m = 4$

2. $p = (-5, -4), m = -8$

3. $p = (-5, 7), m = -3$

4. $p = (2, 4), m = -1$

5. $p = (-1, 2), m = 1$

6. $p = (6, -8), m = 2$

7. $p = (1, -1), m = \frac{1}{2}$

8. $p = (-4, -1), m = -\frac{1}{2}$

9. $p = (6, 9), m = 5$

10. $p = (-1, 2), m = -1$

11. $p = (-2, 9), m = 3$

12. $p = (-2, -1), m = 2$

13. $p = (3, 10), m = -5$

14. $p = (7, 8), m = 4$

© Carson-Dellosa • CD-104632

Name _____

HSA-REI.D.12

Graphing Linear Inequalities

Graph the line $y > 2x + 3$.

1. $m = \frac{2}{1}$ $\quad b = 3$
2. If $>$ or $<$, connect the points with a dotted line.
3. If $\geq$ or $\leq$, connect the points with a solid line.

   The coordinate plane is now divided into 2 regions.
4. Test any (x, y) on each side of the line in the original inequality.

$y > 2x + 3$
Test point A $(^{-}1, 4)$.
Is $4 > 2(^{-}1) + 3$?
$4 > {}^{-}2 + 3$
$4 > 1 \longrightarrow$ true
(Shade this region)
Test point A.

$y > 2x + 3$
Test point B $(0, 0)$.
Is $0 > (0) + 3$?
$0 > 0 + 3$
$0 > 3 \longrightarrow$ false
(Do not shade this region)
Test point B.

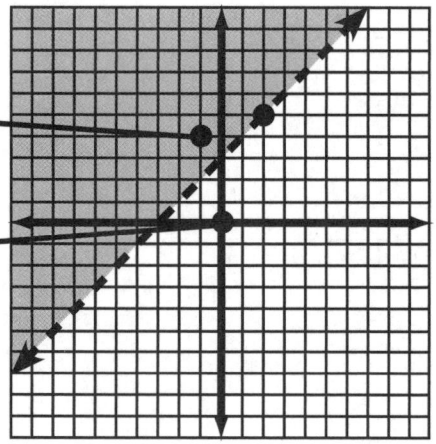

Graph the solution set.

1. $x + 4y > 8$

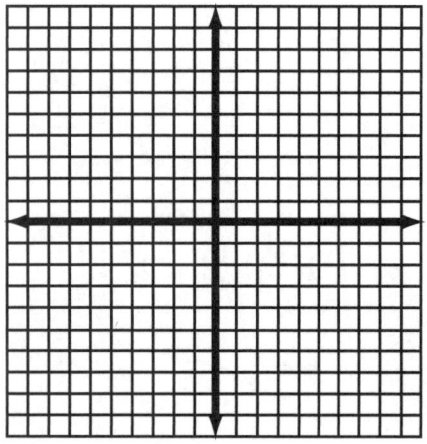

2. $^{-}2x + 2y \geq 10$

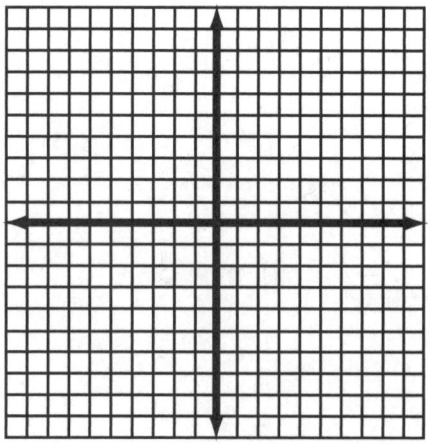

© Carson-Dellosa • CD-104632

Name _____

HSA-REI.D.12

Graphing Linear Inequalities

Graph the solution set.

1. $3x + 2y \geq 6$

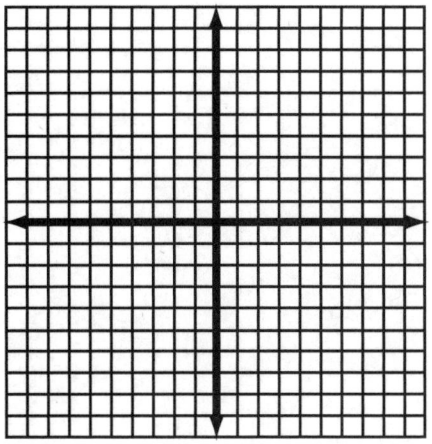

2. $2x + y < 4$

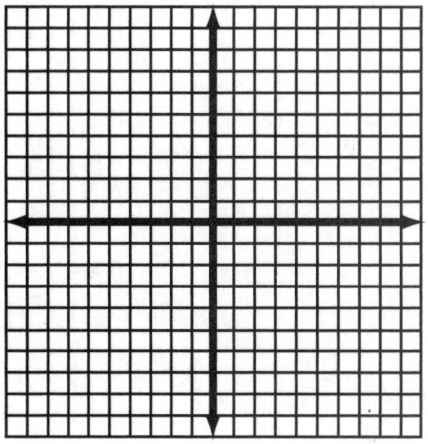

3. $6x - 3y > 15$

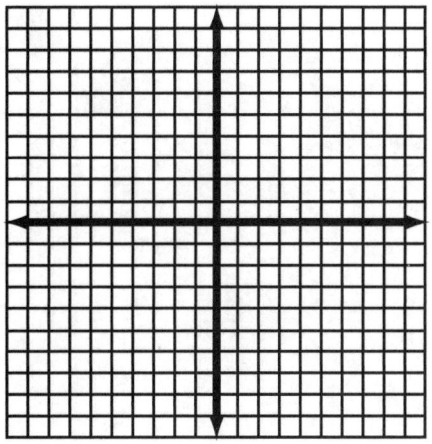

4. $2x + 3y \geq 6$

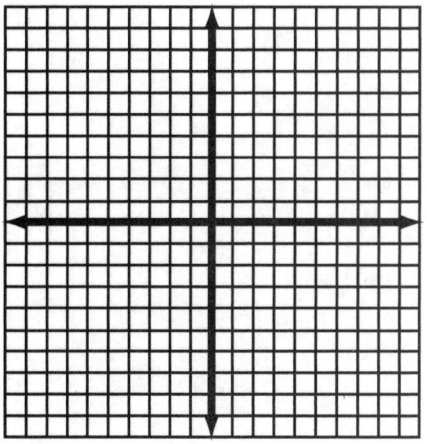

5. $3x - 4y \leq 12$

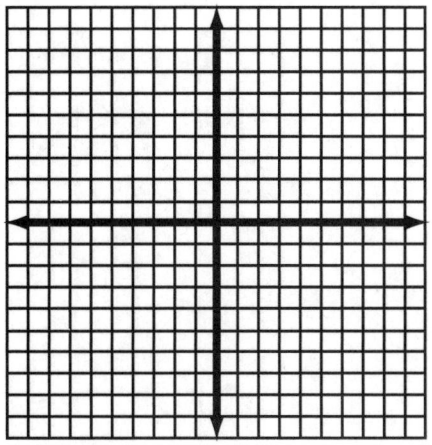

6. $4x + 2y < 6$

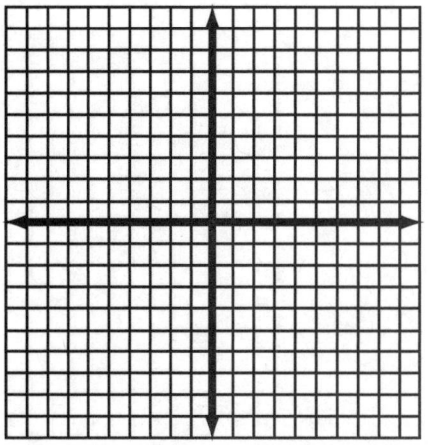

© Carson-Dellosa • CD-104632

Name _____

HSA-REI.D.12

Graphing Linear Inequalities

Graph the solution set.

1. $x + 2y < 0$

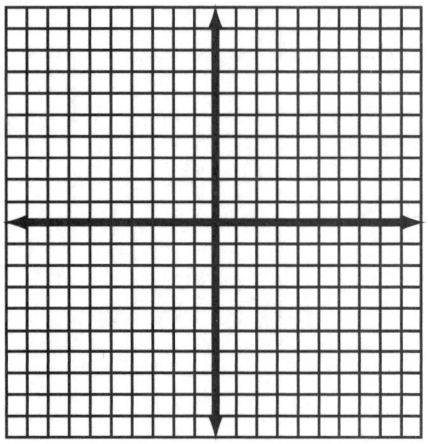

2. $5x - 2y \leq 10$

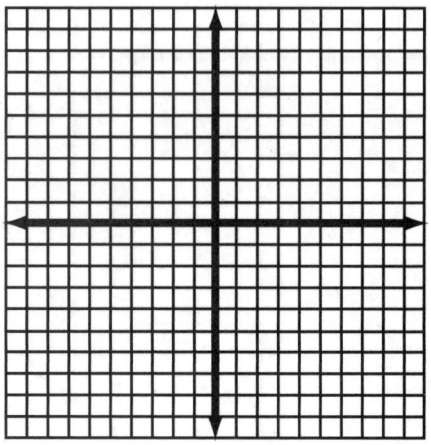

3. $6x - 3y > 18$

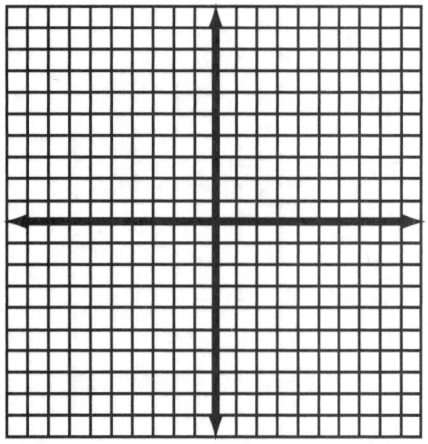

4. $2x - 5y < 10$

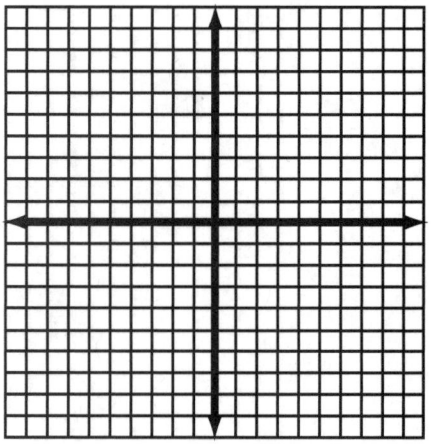

5. $^{-}5x + 5y \leq 15$

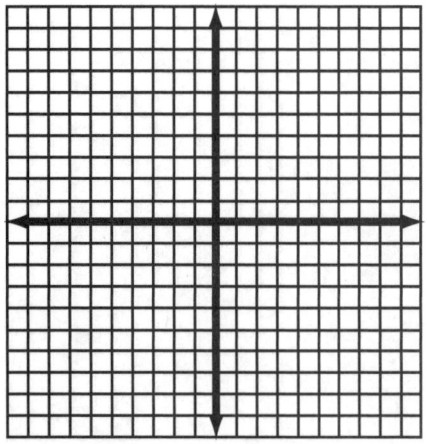

6. $y + 6 > 0$

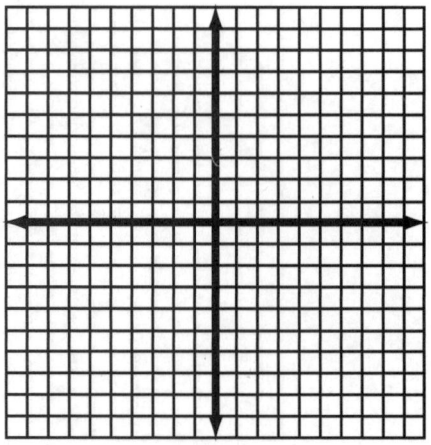

© Carson-Dellosa • CD-104632

Name _____

8.EE.C.8b, HSA-REI.C.6, HSA-REI.D.11

Solving Systems of Equations by Graphing

When two or more equations are considered together, it is called a **system of equations**. The following example is a system of two linear equations in two variables.

$x + 2y = 4$

$2x + y = {}^{-}1$

The graphs of these equations are straight lines.

An ordered pair that is a solution of each equation of the system is a solution of the system of equations in two variables. The solution of a system of linear equations can be found by graphing the lines of the system. The solution of the system of equations is the point where the lines intersect.

Solve by graphing:

$x + 2y = 4$

$2x + y = {}^{-}1$

Graph each line and find the point of intersection.

The solution is $({}^{-}2, 3)$ because the ordered pair lies on each line.

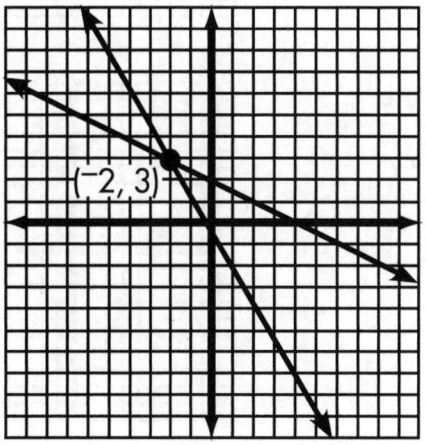

Solve by graphing.

1. $x + y = 7$

   $3x - y = {}^{-}3$

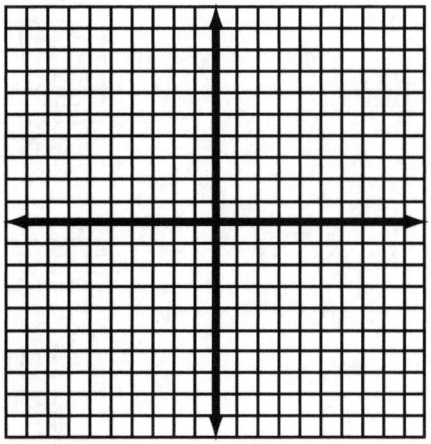

2. $x + y = 4$

   $x - y = 6$

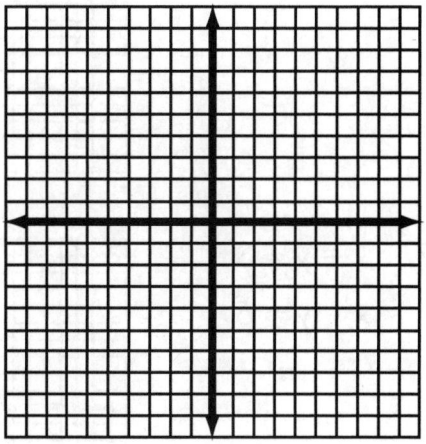

Name _____

8.EE.C.8b, HSA-REI.C.6, HSA-REI.D.11

Solving Systems of Equations by Graphing

Solve by graphing.

1. $3x - 4y = 12$
   $2x + 4y = {}^{-}12$

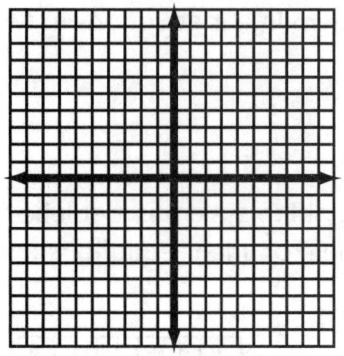

2. $4x - 2y = 8$
   $y = 2$

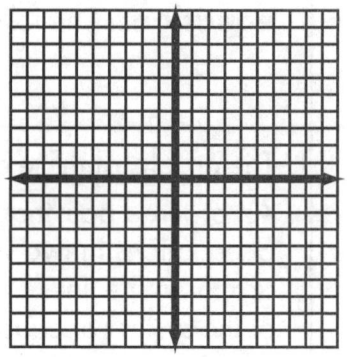

3. $x - y = 1$
   $x + 2y = 10$

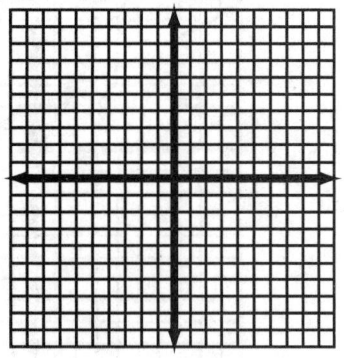

4. $x = 5$
   $y = {}^{-}1$

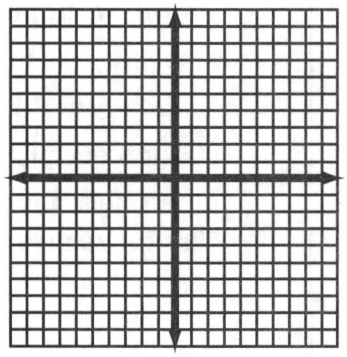

5. $y = 2x + 4$
   $y = x + 6$

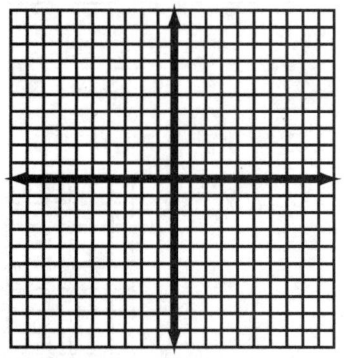

6. $x - y = {}^{-}4$
   $3x - y = {}^{-}12$

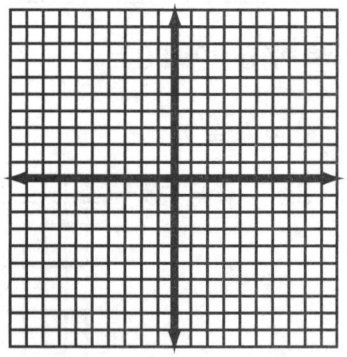

7. $x + y = 3$
   $x - y = 5$

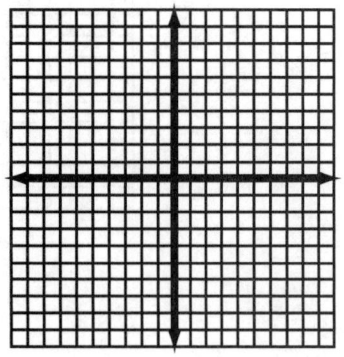

8. $x = 4$
   $6x - 2y = 4$

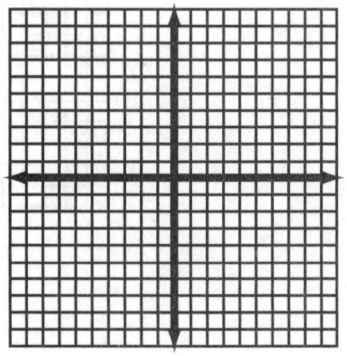

© Carson-Dellosa • CD-104632

Name _____

8.EE.C.8b, HSA-REI.C.6, HSA-REI.D.11

Solving Systems of Equations by Graphing

Solve by graphing.

1. $y = x - 3$
   $x = 2y + 6$

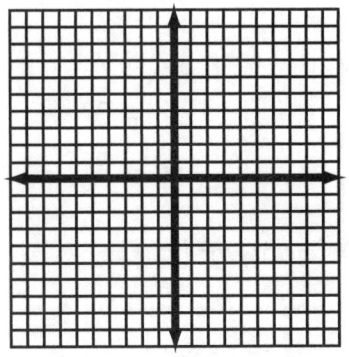

2. $-2x + y = 0$
   $6x + y = -16$

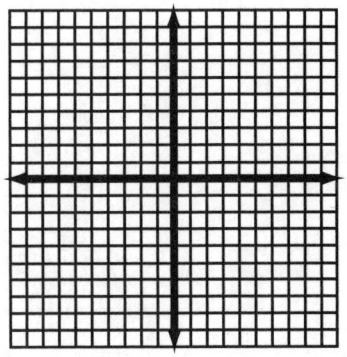

3. $y = 3x + 5$
   $y = 1 - x$

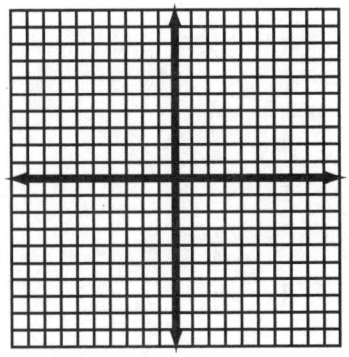

4. $y = x - 5$
   $y + 2x = 1$

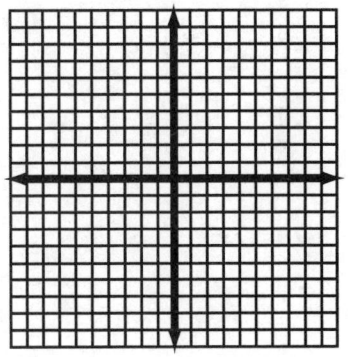

5. $x - 4y = 2$
   $x + 2y = 8$

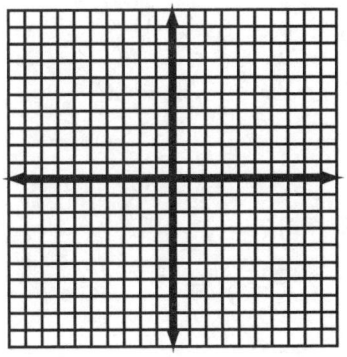

6. $x + y = 6$
   $3x - y = 2$

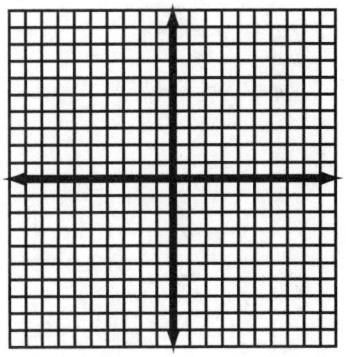

7. $2x + y = 6$
   $3x + y = 12$

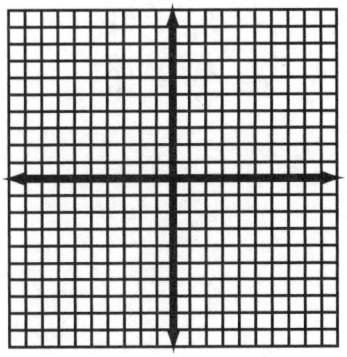

8. $y = x + 3$
   $2y = 3x + 1$

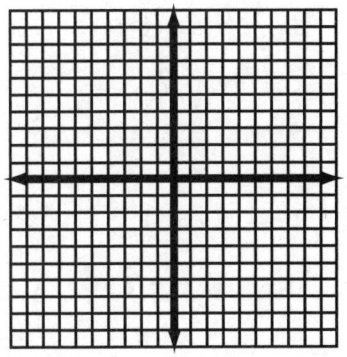

© Carson-Dellosa • CD-104632

Name _____

8.EE.C.8b, HSA-REI.C.6

Solving Systems of Equations

$$\begin{array}{r} 2x + 3y = 7 \\ + \quad x - 3y = {}^{-}1 \\ \hline 3x \qquad = 6 \\ \frac{3x}{3} = \frac{6}{3} \\ \mathbf{x = 2} \end{array}$$

Substitute 2 for x in $2x + 3y = 7$.

$$\begin{array}{r} 2x + 3y = 7 \\ 2(2) + 3y = 7 \\ 4 + 3y = 7 \\ 3y = 3 \\ \mathbf{y = 1} \end{array}$$

answer (2, 1)

Solve.

1. $2x - 5 = y$
   $x - 7 = {}^{-}y$

2. $x + 4y = 2$
   ${}^{-}x + y = 8$

3. $y = 2x - 2$
   ${}^{-}y = x$

4. $3x + y = 8$
   $3x - y = 4$

5. $2x - y = 6$
   $3x + y = 4$

6. $y = 5x + 1$
   $2y = {}^{-}5x + 2$

7. $x + y = 7$
   $x - y = 3$

8. $3x + y = 5$
   $x - y = 7$

9. $3x - 4y = 14$
   $x + 4y = 2$

10. $5x - 3y = {}^{-}1$
    $4x + 3y = 10$

11. $8x - 3y = 1$
    ${}^{-}8x + 5y = 9$

12. $3y - 4x = 5$
    $y + 4x = 7$

© Carson-Dellosa • CD-104632

Name _____

8.EE.C.8b, HSA-REI.C.6

Solving Systems of Equations

1. Solve.

$6x + 5y = 6$

$6x - 3y = 6$

2. $\left.\begin{array}{r} 6x + 5y = 6 \\ (^{-}1)(6x - 3y = 6) \end{array}\right\}$ Multiply to create the additive inverse.

3. $\left.\begin{array}{r} 6x + 5y = 6 \\ {}^{-}6x + 3y = {}^{-}6 \end{array}\right\}$ Use the addition method.

$8y = 0$

$y = 0$

4. $\left.\begin{array}{r} 6x + 5y = 6 \\ 6x + 5(0) = 6 \end{array}\right\}$ Substitute.

$6x = 6$

$x = 1$

Solve.

1. $3x + 8y = 8$
   $2x + y = 1$

2. $3x - 4y = 4$
   $x - y = 1$

3. $4x - 4y = 16$
   $2x + 2y = 4$

4. $2x - 5y = 21$
   $x + y = 7$

5. $4x + 8y = 20$
   $x - y = 2$

6. $6x + 4y = {}^{-}22$
   $3x + 6y = {}^{-}7$

7. $2x + y = 3$
   $x + 3y = 4$

8. $3x + 3y = {}^{-}3$
   $x - y = {}^{-}5$

9. $4x + 2y = 10$
   $x + 2y = 1$

10. $3x + 5y = {}^{-}8$
    $x + 7y = {}^{-}8$

11. $x - y = 7$
    $x + 2y = 1$

12. $2x - 8y = 6$
    ${}^{-}x + 7y = 9$

© Carson-Dellosa • CD-104632

Name _____

8.EE.C.8b, HSA-REI.C.6

Solving Systems of Equations

Solve.

1. $y = 3 - 2x$
   $y = 2 - 3x$

2. $x + y = 5$
   $x = y + 7$

3. $x - y = 1$
   $2x + y = 8$

4. $3x - y = 9$
   $y = x + 5$

5. $3x + 4y = 26$
   $^{-}2x + y = 1$

6. $y = 2x + 3$
   $y = 4x + 4$

7. $2x + 7y = 8$
   $x + 5y = 7$

8. $y = 4x + 4$
   $y = 2x + 8$

9. $x + 3y = 17$
   $2x + 3y = 22$

10. $4x - 7y = 9$
    $y = x - 3$

11. $8x - 5y = 9$
    $y = 2x - 4$

12. $2x + 4y = {}^{-}2$
    $3x + y = 7$

13. $3x + y = 5$
    $2x + 3y = 8$

14. $2x + 6y = 24$
    $x - 4y = {}^{-}2$

© Carson-Dellosa • CD-104632

Name _____ HSN-RN.A.2

Operations with Radicals

Simplify: $\sqrt{49x^4y^8} = 7\sqrt{x^4y^8} = 7x^2y^4$

Multiply: $\sqrt{3a} \bullet \sqrt{4a} = \sqrt{12a^2} = \sqrt{3} \bullet 4 \bullet a^2 = 2a\sqrt{3}$

Divide: $\sqrt{\frac{18}{2}} = \sqrt{9} = 3$ $\sqrt{\frac{9}{25}} = \frac{\sqrt{9}}{\sqrt{25}} = \frac{3}{5}$

Add or subtract: $2\sqrt{y} + 3\sqrt{y} + \sqrt{y} = 6\sqrt{y}$ $\sqrt{4x} + 3\sqrt{x} = 2\sqrt{x} + 3\sqrt{x} = 3\sqrt{x}$

Simplify.

1. $\sqrt{x^2y^{10}} =$

2. $\sqrt{27x^8} =$

3. $\sqrt{x^{16}} =$

4. $\sqrt{125b^{15}} =$

5. $\sqrt{x^2y^4} \bullet 2\sqrt{xy} =$

6. $\sqrt{9} \bullet \sqrt{32} =$

7. $3\sqrt{5} \bullet 2\sqrt{4} =$

8. $2\sqrt{4x^3y} \bullet y\sqrt{x^5y^7} =$

9. $\sqrt{\frac{36}{9}} =$

10. $\sqrt{\frac{27x}{3x}} =$

11. $3\sqrt{x^3} - 4\sqrt{x^3} =$

12. $3\sqrt{2y} + 2\sqrt{2y} =$

13. $2\sqrt{y} - 4\sqrt{y} =$

14. $3y\sqrt{2y} - y\sqrt{2y} =$

© Carson-Dellosa • CD-104632

Name _____

HSN-RN.A.2

Operations with Radicals

Simplify.

1. $\sqrt{x^{14}y^6} =$

2. $\sqrt{169y^{12}} =$

3. $\sqrt{16x^4} =$

4. $\sqrt{8x^3} =$

5. $\sqrt{81x^6} =$

6. $\sqrt{25x^6} =$

7. $4\sqrt{9x^3} \bullet 3\sqrt{4x} =$

8. $x\sqrt{5x^3y} \bullet x\sqrt{5x^2y} =$

9. $2\sqrt{9x^2} \bullet 2\sqrt{4x^2} =$

10. $6\sqrt{9xy} \bullet 4\sqrt{2xy} =$

11. $2\sqrt{4x^3y} \bullet 3\sqrt{3a^2b^2} =$

12. $\sqrt{\frac{x^2}{25}} =$

13. $\sqrt{\frac{8x^3}{2x}} =$

14. $\sqrt{\frac{18x^3}{2x}} =$

15. $x\sqrt{27} + x\sqrt{12} =$

16. $3\sqrt{6x} + 5\sqrt{6x} =$

17. $4\sqrt{2x^3} + 3\sqrt{2x^3} =$

18. $2\sqrt{50} - 4\sqrt{8} - 3\sqrt{72} =$

© Carson-Dellosa • CD-104632

Name _____

HSN-RN.A.2

Operations with Radicals

Simplify.

1. $\sqrt{x^9y^9} =$

2. $\sqrt{54x^8} =$

3. $\sqrt{9a^4b^8} =$

4. $\sqrt{49x^4y^2} =$

5. $5\sqrt{2x^6y} \bullet 3\sqrt{3x^3y^5} =$

6. $4\sqrt{8a^6b} \bullet 4\sqrt{8a^4b^4} =$

7. $3\sqrt{2x^3} \bullet 3\sqrt{3x^2y^2} =$

8. $5\sqrt{4a} \bullet 2\sqrt{6a} =$

9. $x\sqrt{3x} \bullet \sqrt{3x^3} =$

10. $\sqrt{2x^4} \bullet \sqrt{10x^2y^2} =$

11. $\sqrt{\frac{9}{64}} =$

12. $\sqrt{\frac{50x^2}{2}} =$

13. $\sqrt{\frac{49x^2}{25x^3}} =$

14. $\sqrt{\frac{12x^2}{60}} =$

15. $\sqrt{\frac{3x^7}{108y^2}} =$

16. $4\sqrt{y^3} - 2\sqrt{y^3} =$

17. $y\sqrt{y^4} - 2y\sqrt{y^4} =$

18. $4x\sqrt{x^3} + 2x\sqrt{x^3} =$

19. $4\sqrt{x} - 2\sqrt{x} - 3\sqrt{x} + 5\sqrt{x} =$

20. $3\sqrt{4x^2y} - 8y\sqrt{y} =$

© Carson-Dellosa • CD-104632

Answer Key

Name _____ (7.NS.A.1d)

Adding Real Numbers

A number line is a great way to model the addition of real numbers.

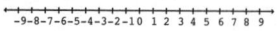

Add 3 + 5. (Start at 3 and move 5 places to the right since 5 is positive.) The answer is 8.
Add 3 + (−5). (Start at 3 and move 5 places to the left since 5 is negative.) The answer is −2.
When adding a number that is positive, move to the right. When adding a number that is negative, move to the left.

Explain how to add the following numbers using a number line, and tell the result.

1. 7 + 8 **Start at 7 and move 8 places to the right. 15**
2. −5 + 9 **Start at −5 and move 9 places to the right. 4**
3. 10 + (−7) **Start at 10 and move 7 places to the left. 3**
4. −11 + (−6) **Start at −11 and move 6 places to the left. −17**

Use the number line to add the numbers.

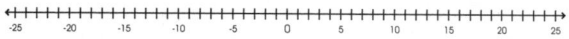

5. 4 + 9 **13**
6. −5 + 13 + (−11) **−3**
7. −7 + (−6) **−13**
8. 14 + 10 + (−3) **21**
9. 0 + (−5) **−5**
10. −6 + (−9) + 10 **−5**
11. −3 + 3 **0**
12. 8 + (−2) + (−4) + 6 **8**
13. 7 + (−12) **−5**
14. −11 + 5 + (−13) + 8 **−11**
15. −11 + 8 + (−1) **−4**
16. 12 + (−6) + (−14) + 17 **9**

© Carson-Dellosa • CD-104632

Name _____ (7.NS.A.1d)

Adding Real Numbers

−7 + 6 = −1

Add.

1. −55 + (−8) + (−4) + 54 = **−13**
2. 1 + (−5) + (−5) + 1 = **−8**
3. 2.7 + (−4.8) = **−2.1**
4. 3.54 + 4.27 + 7.43 = **15.24**
5. 10 + 7 + (−7) + (−10) = **0**
6. 16 + 21 + (−3) + 7 = **41**
7. 10 + 7 + (−16) + 9 + (−30) = **−20**
8. 5.8 + 8.4 = **14.2**
9. 2.76 + (−6.56) + (−9.72) = **−13.52**
10. 8 + (−7) = **1**
11. $2\frac{3}{5} + 4\,3\frac{3}{7} =$ **$7\frac{1}{35}$**
12. $-8\frac{3}{5} + 3\frac{3}{7} =$ **$-5\frac{6}{35}$**
13. 2 + 5 + (−3) = **4**
14. $-5\frac{3}{4} + (-2\frac{3}{4}) + 8 =$ **$-\frac{1}{2}$**
15. 7.3 + (3.9) = **11.2**
16. −21 + 12 + (−1) + (−17) = **−27**
17. 7.867 + (−5.329) = **2.538**
18. $-2\frac{3}{5} + (-5\frac{3}{7}) + 3 =$ **$-5\frac{1}{35}$**
19. 3 + 12 + (−13) + 36 = **38**
20. $-3\frac{1}{6} + (-9\frac{3}{12}) + 6 =$ **$-6\frac{5}{12}$**

© Carson-Dellosa • CD-104632

Name _____ (7.NS.A.1d)

Adding Real Numbers

Add.

1. $2\frac{3}{5} + (-3\frac{2}{5}) + -6 =$ **$-6\frac{4}{5}$**
2. 21 + 9 + (−6) + 7 = **31**
3. 12 + (−9) + 17 = **20**
4. 2.54 + −5.87 + −32.65 = **−35.98**
5. 1.45 + 2.65 + (−9.43) = **−5.33**
6. 21 + 3 + (−13) + 22 = **33**
7. 3 + (−3) + 4 + (−5) = **−1**
8. 3.3 + (−3.4) + 5.5 = **5.4**
9. 3.6 + (−2.5) + −5.5 = **−4.4**
10. −0.6 + (−0.56) + 3 = **1.84**
11. $3\frac{5}{8} + (-1\frac{3}{4}) + 2 =$ **$3\frac{7}{8}$**
12. 4.524 + 7.342 = **11.866**
13. $-7\frac{2}{4} + 2\frac{3}{4} =$ **−4 3/4**
14. 34 + (−13) + 18 + 0 + 34 = **73**
15. 8.43 + (−10.98) + (−3.23) = **−5.78**
16. 2.54 + (−5.21) + (−6.34) = **−9.01**
17. $-2\frac{1}{3} + (-5\frac{7}{10}) + (-7) =$ **$-15\frac{1}{30}$**
18. $-1\frac{2}{3} + (-3\frac{3}{5}) + 4 =$ **$-1\frac{4}{15}$**
19. $2\frac{1}{2} + 6\frac{1}{2} =$ **9**
20. $4\frac{3}{5} + (-3\frac{2}{5}) + (-8) =$ **$-6\frac{4}{5}$**

© Carson-Dellosa • CD-104632

Name _____ (7.NS.A.1c, 7.NS.A.1d)

Subtracting Real Numbers

For all real numbers *a* and *b*, *a* − *b* = *a* + (−*b*). Simply stated: To subtract a number, add its opposite.

7 − 10 = 7 + (−10) = −3	5 − 8 + 3 − 1 = 5 + (−8) + 3 + (−1) = 5 + 3 + (−8) + (−1) = 8 + (−9) = −1 (group the numbers with like signs)	−4 − (−12) = −4 + 12 = 8

Change each problem into an addition problem.

1. 7 − 9 **7 + (−9)**
2. −6 − (−4) **−6 + 4**
3. −11 − 5 **−11 + (−5)**
4. 12 − (−15) **12 + 15**
5. 8 − 3 **8 + (−3)**
6. 22 − 5 **22 + (−5)**
7. 4 − 11 **4 + (−11)**
8. −4 − (−9) **−4 + 9**

Subtract.

9. 9 − 11 **−2**
10. 0 − (−12) **12**
11. −5 − 4 **−9**
12. 6 − (−6) **12**
13. −1 − (−1) **0**
14. −7 − 6 **−13**
15. 3 − (−5) **8**
16. 17 − 23 **−6**

© Carson-Dellosa • CD-104632

© Carson-Dellosa • CD-104632

Answer Key

Name _____ 7.NS.A.1c, 7.NS.A.1d

Subtracting Real Numbers

$10 - (^-4) = 10 + 4 = 14$

Subtract.

1. $9 - (^-32) =$ **41**
2. $^-99 - (^-42) =$ **-57**
3. $4 - (^-8) =$ **12**
4. $0 - 21 =$ **-21**
5. $45 - 301 =$ **-256**
6. $^-19 - 8 =$ **-27**
7. $^-43 - 6 =$ **-49**
8. $9 - (^-2) - 8 - 7 =$ **-4**
9. $35 - 67 - 85 - 21 - 12 =$ **-150**
10. $12 - 7 - (^-16) - 9 - (^-34) =$ **46**
11. $18 - (^-13) =$ **31**
12. $121 - 45 =$ **76**
13. $-\frac{4}{7} - \frac{1}{3} - (\frac{2}{3}) =$ **$-1\frac{4}{7}$**
14. $3.434 - 7.294 =$ **-3.86**
15. $8 - 2.8 =$ **5.2**
16. $8 - (^-14) =$ **22**
17. $3.9 - 4.9 =$ **-1**
18. $^-7 - (^-3) =$ **-4**
19. $2.19 - 7.8 - 8.31 =$ **-13.92**
20. $38 - 39 - (^-13) =$ **12**

© Carson-Dellosa • CD-104632

Name _____ 7.NS.A.1c, 7.NS.A.1d

Subtracting Real Numbers

Subtract.

1. $^-9 - (^-5) =$ **-4**
2. $321 - (^-34) =$ **355**
3. $\frac{2}{3} - \frac{4}{5} =$ **$-\frac{2}{15}$**
4. $\frac{3}{5} - \frac{7}{8} =$ **$-\frac{11}{40}$**
5. $5.34 - 9.9 - 3.65 =$ **-8.21**
6. $9.432 + 4.348 - 44.938 =$ **-31.158**
7. $245 - 32 - (^-36) =$ **249**
8. $44 - 35 - 34 - 32 =$ **-57**
9. $8 - (^-5) - 7 - 9 =$ **-3**
10. $43 - 88 - 35 - 21 =$ **-101**
11. $-\frac{2}{5} - \frac{3}{4} - (-\frac{4}{5}) =$ **$-\frac{7}{20}$**
12. $^-45 - 5 =$ **-50**
13. $-\frac{2}{3} - \frac{1}{3} - (-\frac{1}{3}) =$ **$-\frac{2}{3}$**
14. $-\frac{4}{5} - \frac{1}{2} - \frac{2}{5} =$ **$-1\frac{7}{10}$**
15. $4 - 12.9 =$ **-8.9**
16. $7 - (^-33) =$ **40**
17. $3.4 - 7.4 =$ **-4**
18. $2.456 - 4.345 - 5.457 =$ **-7.346**
19. $23 - (^-21) =$ **44**
20. $4.346 - 0.4537 =$ **3.8923**

© Carson-Dellosa • CD-104632

Name _____ 7.NS.A.2a, 7.NS.A.2c

Multiplying Real Numbers

The **property of zero for multiplication** states that for all real numbers a, $a \cdot 0 = 0$ and $0 \cdot a = 0$. Simply stated, any real number multiplied by 0 is 0. For example, $0 \cdot 20 = 0$.

To multiply two real numbers with the same signs:
1. Multiply their absolute values.
2. The sign of their product is positive.

positive • positive = positive
(+) (+) (+)
$3 \cdot 12 = 36$

negative • negative = positive
(−) (−) (+)
$^-7 \cdot {}^-8 = 56$

To multiply two real numbers with different signs:
1. Multiply their absolute values.
2. The sign of their product is negative.

negative • positive = negative
(−) (+) (−)
$^-2 \cdot 5 = {}^-10$

positive • negative = negative
(+) (−) (−)
$4 \cdot {}^-8 = {}^-32$

Write the sign of the product for each number.

1. $(^-10)4$ **−**
2. $8(^-1)$ **−**
3. $(^-2)(^-3)$ **+**
4. $(7)(5)(^-3)$ **−**
5. $5(6)$ **+**
6. $(^-2)(^-7)$ **+**
7. $(^-12)(^-4)(^-1)$ **−**
8. $(^-6)(4)(^-2)$ **+**

Multiply to find each product.

9. $4(6)(^-1)$ **-24**
10. $(^-1)(^-4)(^-3)$ **-12**
11. $(-\frac{1}{2})(2)$ **-1**
12. $(7)(^-3)(0)$ **0**
13. $(5)(3)$ **15**
14. $(-\frac{1}{8})(^-16)(4)$ **8**
15. $(^-9)(^-4)$ **36**
16. $(^-7)(7)$ **-49**

© Carson-Dellosa • CD-104632

Name _____ 7.NS.A.2a, 7.NS.A.2c

Multiplying Real Numbers

$(^-2)(^-3) = 6$

Multiply.

1. $4 \cdot 9 =$ **36**
2. $^-4 \cdot 12 =$ **-48**
3. $(-\frac{5}{9})(8.8) =$ **$4.\overline{8}$**
4. $(^-3)(0) =$ **0**
5. $(^-3)(^-9) =$ **27**
6. $6(23) =$ **138**
7. $(12)(^-3)(4) =$ **-144**
8. $(^-5)(^-5)(^-5) =$ **-125**
9. $(5)(2)(^-1) =$ **-10**
10. $(7)(^-9)(^-12) =$ **756**
11. $(-\frac{2}{3})(^-1.6) =$ **$1.0\overline{6}$**
12. $^-7(^-7) =$ **49**
13. $(54.2)(^-3.55) =$ **-192.41**
14. $(2.22)(^-1.11) =$ **-2.4642**
15. $(7.44)(3.2)(4.3) =$ **102.3744**
16. $(2.4)(^-1.4) =$ **-3.36**
17. $(-\frac{3}{5})(\frac{3}{5}) =$ **$-\frac{9}{25}$**
18. $(-\frac{4}{5})(2.2) =$ **-1.76**
19. $^-8 \cdot 12 =$ **-96**
20. $(0)(2)(^-213) =$ **0**

© Carson-Dellosa • CD-104632

© Carson-Dellosa • CD-104632

Answer Key

Name _____ (7.NS.A.2a, 7.NS.A.2c)

Multiplying Real Numbers

Substitute and multiply.

$x = {}^{-}3, y = {}^{-}5, z = 0$

1. xy **15**
2. ${}^{-}3yz$ **0**
3. xyz **0**
4. $4x({}^{-}2y)$ **−120**
5. ${}^{-}3(10xy)$ **−450**
6. $2y(5y)$ **250**

$x = {}^{-}2.4, y = 3.1, z = {}^{-}4.8$

7. $4xz$ **46.08**
8. $yz({}^{-}z)$ **−71.424**
9. ${}^{-}5(xyz)$ **−178.56**
10. ${}^{-}x({}^{-}6z)$ **69.12**
11. $(xy)({}^{-}3x)$ **−53.568**
12. $(3yz)({}^{-}7x)$ **−749.952**

$x = \frac{1}{2}, y = -\frac{2}{3}, z = \frac{3}{5}$

13. $8(xy)$ **$-2\frac{2}{3}$**
14. xyz **$-\frac{1}{5}$**
15. ${}^{-}12x({}^{-}5yz)$ **−12**
16. $(xy)(3xz)$ **$-\frac{3}{10}$**
17. $({}^{-}2z)(3y)(6x)$ **$7\frac{1}{5}$**
18. ${}^{-}z(10xyz)$ **$1\frac{1}{5}$**

Name _____ (7.NS.A.2b, 7.NS.A.2c)

Dividing Real Numbers

The **multiplicative inverse property** states that for each nonzero a, there is exactly one number $\frac{1}{a}$ such that: $a \cdot \frac{1}{a} = 1$ and $\frac{1}{a} \cdot a = 1$.

The number $\frac{1}{a}$ is called the reciprocal of a.

For example, the reciprocal of 9 is $\frac{1}{9}$ and the reciprocal of $\frac{1}{2}$ is 2. Simply flip the number to find its reciprocal.

To divide the number a by the number b, multiply a by the reciprocal of b.

$a \div b = a \cdot \frac{1}{b}$ (The result will be the quotient of a and b.)

$12 \div 4 = 12 \cdot \frac{1}{4} = 3$ $\quad 4x \div \frac{1}{4} = 4x \cdot 4 = 16x$ $\quad 15 \div \frac{3}{2} = 15 \cdot \frac{2}{3} = 10$

To divide two nonzero real numbers:

1. The quotient is positive if both numbers have the same sign.
2. The quotient is negative if both numbers have different signs.

${}^{-}15 \div 5 = {}^{-}15 \cdot \frac{1}{5} = {}^{-}3$ $\quad 3 \div \frac{1}{4} = 3 \cdot 4 = 12$ $\quad 1 \div -\frac{4}{5} = 1 \cdot -\frac{5}{4} = -\frac{5}{4}$

Write the reciprocal of each number. Write none if it does not exist.

1. 2 **$\frac{1}{2}$**
2. 1 **1**
3. 0 **none**
4. $-\frac{1}{9}$ **−9**
5. $-\frac{1}{4}$ **−4**
6. $\frac{4}{3}$ **$\frac{3}{4}$**
7. ${}^{-}8$ **$-\frac{1}{8}$**
8. 10 **$\frac{1}{10}$**

Divide.

9. $\frac{9}{3}$ **3**
10. $\frac{{}^{-}28}{7}$ **−4**
11. ${}^{-}15 \div 3$ **−5**
12. $\frac{{}^{-}7}{8} \div \frac{1}{8}$ **−7**
13. $36 \div ({}^{-}4)$ **−9**
14. $\frac{{}^{-}8}{{}^{-}8}$ **1**
15. $44 \div ({}^{-}11)$ **−4**
16. $0 \div \frac{4}{5}$ **0**

Name _____ (7.NS.A.2b, 7.NS.A.2c)

Dividing Real Numbers

$9 \div 4.5 = 2$

Divide.

1. $\frac{49}{7} =$ **7**
2. $90 \div 15 =$ **6**
3. $({}^{-}12) \div (9.9) =$ **$-1.\overline{21}$**
4. $(-\frac{2}{3}) \div ({}^{-}18) =$ **$-\frac{1}{27}$**
5. ${}^{-}42 \div 7 =$ **−6**
6. $45 \div ({}^{-}8) =$ **−5.625**
7. $\frac{{}^{-}36}{4} =$ **−9**
8. $(-\frac{3}{5}) \div (\frac{3}{5}) =$ **−1**
9. ${}^{-}72 \div (9) =$ **−8**
10. ${}^{-}21 \div ({}^{-}9) =$ **$2.\overline{3}$**
11. $\frac{102}{17} =$ **6**
12. $0 \div ({}^{-}8) =$ **0**
13. $\frac{95}{5} =$ **19**
14. $\frac{63}{{}^{-}9} =$ **−7**
15. $({}^{-}3.4) \div ({}^{-}9.99) =$ **$0.\overline{340}$**
16. ${}^{-}50 \div [40 \div ({}^{-}20)] =$ **25**
17. $(-\frac{4}{6}) \div (36) =$ **$-\frac{1}{54}$**
18. $56 \div ({}^{-}28 \div 7) =$ **−14**
19. $({}^{-}38 \div 19) \div ({}^{-}2) =$ **1**
20. ${}^{-}45 \div [{}^{-}20 \div ({}^{-}4)] =$ **−9**

Name _____ (7.NS.A.2b, 7.NS.A.2c)

Dividing Real Numbers

Substitute and divide.

$x = {}^{-}24, y = 6, z = {}^{-}8$

1. $\frac{x}{y}$ **−4**
2. $\frac{yz}{{}^{-}x}$ **−2**
3. $\frac{{}^{-}z}{xy}$ **−18**
4. $\frac{4z}{5y}$ **$-1\frac{1}{15}$**
5. $\frac{{}^{-}xy}{{}^{-}yz}$ **3**
6. $\frac{2y}{{}^{-}5x}$ **$\frac{1}{10}$**

$w = 2.8, x = 1.4, y = {}^{-}8.4, z = {}^{-}11.2$

7. $(w) \div ({}^{-}x)$ **2**
8. $(z) \div ({}^{-}y)$ **−1.3**
9. $\frac{x}{({}^{-}w)}$ **−0.5**
10. $({}^{-}z) \div ({}^{-}w)$ **−4**
11. $\frac{{}^{-}3y}{2x}$ **9**
12. $(z) \div ({}^{-}x)$ **−8**

$x = -\frac{1}{4}, y = -\frac{1}{3}, z = \frac{4}{5}$

13. $(z) \div (x)$ **$-3\frac{1}{5}$**
14. $\frac{x}{{}^{-}z}$ **$\frac{5}{16}$**
15. $\frac{({}^{-}2x)}{(y)}$ **$-1\frac{1}{2}$**
16. $(y) \div ({}^{-}xz)$ **$-1\frac{2}{3}$**
17. $({}^{-}z) \div (y) \div ({}^{-}x)$ **$9\frac{3}{5}$**
18. $\frac{{}^{-}z}{xy}$ **$-9\frac{3}{5}$**

Answer Key

Name _____ (5.OA.A.1, 6.EE.A.2c)

Order of Operations

When solving an equation, be sure to follow the **order of operations**.

1. Parentheses
2. Exponents
3. Multiplication & Division
4. Addition & Subtraction

$14 - 54 \div 6 = 14 - 9 = 5$

Solve.

1. $3 \times 15 \div 5 =$ **9**
2. $35 \div 5 - 9 =$ **−2**
3. $3 + 2 \times 4 =$ **11**
4. $5 \times 2 \times 8 =$ **80**
5. $6 - 40 \div 8 =$ **1**
6. $12 - 30 \div 6 =$ **7**
7. $32 \div 4 \times 3 =$ **24**
8. $8 + 3 \times 2 =$ **14**
9. $4 + 12 \div 2 =$ **10**
10. $9 + 20 \div 5 =$ **13**
11. $15 - 75 \div 5 =$ **0**
12. $9 - 3 + 6 =$ **12**
13. $2 \times 8 \div 4 =$ **4**
14. $3 + 3 - 2 =$ **4**

© Carson-Dellosa • CD-104632

Name _____ (5.OA.A.1, 6.EE.A.1, 6.EE.A.2c)

Order of Operations

Remember to follow the **order of operations**.

1. Parentheses
2. Exponents
3. Multiplication & Division
4. Addition & Subtraction

$(3^3 + 6 \times 5) - 2 = (27 + 6 \times 5) - 2 = (27 + 30) - 2 = 55$

Solve.

1. $(3^2 + 2 \times 3) \div 5 =$ **3**
2. $5^2 - 4^2 + 2 =$ **11**
3. $(4 + 2)^2 =$ **36**
4. $(11 - 8)^3 =$ **27**
5. $2(7 + 2) =$ **18**
6. $(9 - 7)^3 - (4 + 3) =$ **1**
7. $(14 - 6)2 =$ **16**
8. $4 + 3(12 - 9) =$ **13**
9. $5^2 - 2^3 =$ **17**
10. $3 \times 8 - (3 \times 2 + 7) =$ **11**
11. $(5^2 - 3 \times 5) \div 2 =$ **5**
12. $7 + 2^2(5 + 2) =$ **35**
13. $3 + 7^2 =$ **52**
14. $(2^2 + 3)^2 - 4 =$ **45**
15. $6 + 7 \times 3 - 9 \times 2 =$ **9**
16. $(2 \times 3) + (21 \div 7) =$ **9**
17. $7^2 - 2(3 \times 3 + 5) =$ **21**
18. $3 + (6 \times 2) =$ **15**

© Carson-Dellosa • CD-104632

Name _____ (5.OA.A.1, 6.EE.A.1, 6.EE.A.2c)

Order of Operations

Solve.

1. $8 - 4 \bullet 5(3 - 2) + 3 =$ **−9**
2. $12 \div (2 - 7) + 7 =$ **4.6**
3. $(14 - 9) + 4 =$ **9**
4. $\frac{3^2 - 5 \bullet 7 - 4^2}{(^-4 - 7 - 12) + 8} =$ **$2\frac{4}{5}$**
5. $9(3 \div 3) + 4(^-5 \bullet 9) \div 3 =$ **−51**
6. $3 - (6 \bullet 6) - 3 \bullet 0 =$ **−33**
7. $36 \div 9 - 8 + 21 \div 3 =$ **3**
8. $5(3 - 8) \bullet 3 + 8 - 3 =$ **−70**
9. $3 \bullet 5 + 9 \bullet 7 =$ **78**
10. $\frac{(5 - 9)^2 + 2}{(7 - 8)^2 \bullet 3^2} =$ **2**
11. $4^2 + 3^2 - 7^2 =$ **−24**
12. $\frac{3^2 - 10}{4^2 - 12} =$ **$\frac{1}{4}$**
13. $8^2 - \frac{26}{(4 + 9)} + 4 =$ **66**
14. $\frac{5 \bullet 7 - (3 + 4)}{^-(2^2) - 2^2 + 3^2} =$ **28**
15. $\frac{4 + 2 \bullet 3 + 4 - 3}{2^2 \bullet 3^2 - 3} =$ **$\frac{1}{3}$**
16. $\frac{3 + 10 - 19 + 32}{3^2 - 1 + 2^2} =$ **$2\frac{1}{6}$**
17. $12 \div [3 + (6 + 3)] =$ **1**
18. $3 \bullet (0 - 7) + 8 \div 2^2 =$ **−19**

© Carson-Dellosa • CD-104632

Name _____ (6.NS.C.5, 6.NS.C.6a, 6.NS.C.7)

Real-Number Operations with Absolute Value

Every real number has an opposite. Opposite numbers are the same distance from 0 on a number line and lie on the opposite sides of 0. The opposite of a positive number is a negative number. The opposite of a negative number is a positive number. The numbers 3 and −3 are opposites. Find their places on the real number line to the right. The additive inverse of a number is the same as the opposite of a number.

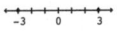

Remember, the opposite of 0 is simply 0 since it is neither positive nor negative.

The symbol $|x|$ is called the absolute value of x. The absolute value of a number is the distance between the number and 0 on a number line. The absolute value of a number, whether positive or negative, is always positive.

$|10| = 10$ $\quad |^-2| = 2$ $\quad |^-102| = 102$ $\quad ^-|12| = ^-12$

Note: The answer to $^-|12|$ is $^-12$ because the absolute value of 12 is 12. But, it is multiplied by a negative, which resulted in $^-12$.

Write the opposite of each number.

1. 5 **−5**
2. −2.6 **2.6**
3. −40 **40**
4. 2.8 **−2.8**

Write a real number to represent each situation.

5. a gain of 12 yards **12**
6. a temperature drop of 8° **−8**
7. a deposit of $89.26 **89.26**
8. a withdrawal of $75 **−75**

Write the absolute value of each number.

9. $|10|$ **10**
10. $|23|$ **23**
11. $|0|$ **0**
12. $|^-42|$ **42**

Simplify.

13. $^-|^-15|$ **−15**
14. $|^-5| \bullet |^-4|$ **20**
15. $^-|4 \bullet 3|$ **−12**
16. $^-|30 \div 6|$ **−5**
17. $|^-8| + |8|$ **16**
18. $|7| - |^-7|$ **0**

© Carson-Dellosa • CD-104632

© Carson-Dellosa • CD-104632

Answer Key

Name _____ 6.NS.C.6a, 6.NS.C.7

Real-Number Operations with Absolute Value

The **absolute value** of a number is its distance from zero.
$-|5 - 11| = -|-6| = -6$ $|-4| \cdot |-3| = 4 \cdot 3 = 12$

Simplify.

1. $|-3| =$ **3**
2. $|-14| =$ **14**
3. $9 + |-4| =$ **13**
4. $-5 \cdot |4| + |5| =$ **−15**
5. $-|4 \cdot 7| =$ **−28**
6. $|21| + 9 =$ **30**
7. $23 + |8| =$ **31**
8. $|12| - |-15| =$ **−3**
9. $7 - |-23| + |-7| =$ **−9**
10. $|-6| \cdot |8| =$ **48**
11. $-|-3 + 7| =$ **−4**
12. $-|-5 + 10| \div |-3| =$ $-\frac{5}{3}$
13. $|-9| + |23| =$ **32**
14. $|-17| \div |-17| =$ **1**
15. $|1| - |0| + 6 =$ **7**
16. $|-67| - |-17| =$ **50**
17. $|4| - |-12 \div 4| =$ **1**
18. $|3 - 13| \div |-5| =$ **2**
19. $|24| \div |-12| =$ **2**
20. $-|9| \cdot |-9| =$ **−81**

© Carson-Dellosa • CD-104632

Name _____ 6.NS.C.6a, 6.NS.C.7

Real-Number Operations with Absolute Value

Simplify.

1. $|-4 + 8| =$ **4**
2. $-|5| \cdot |-7| =$ **−35**
3. $12 \div |-3 + 7| =$ **3**
4. $-|11| - |-20| =$ **−31**
5. $|6 - 18| + |-17 + 9| =$ **20**
6. $|-27| \div |-3| \cdot -|-2| =$ **−18**
7. $-48 \div |-2 \cdot 6| =$ **−4**
8. $|-83| - |-38| =$ **45**
9. $|24| \cdot |-4 + 1| =$ **72**
10. $10 - |-13| + |-7| =$ **4**
11. $|72 \div 8| \cdot |-11| =$ **99**
12. $-|16 - 21| - |-7 + 4| =$ **−8**
13. $23 + |-45 \div 9| - |52| =$ **−24**
14. $-|-4 - 15| =$ **−19**
15. $|-83| + |-16| =$ **99**
16. $-|4| \cdot |-4| =$ **−16**
17. $|-7 + 15| \div |-20 - 12| =$ $\frac{1}{4}$
18. $-41 + |-29| \cdot |0| =$ **0**
19. $|-15| \div |3| - |45| =$ **−40**
20. $|22 \div 11| \cdot |-32 \div 8| =$ **8**

© Carson-Dellosa • CD-104632

Name _____ 6.EE.A.2b, 6.EE.A.3

Combining Like Terms

The expression $3x^2 + 2x + 1$ has three terms, $3x^2$, $2x$, and 1. A **term** is either a single number, a variable, or numbers and variables multiplied together. A term in an expression without a variable is called a **constant**, as 1 is above. For terms to be considered "like" terms, they must have the same variable and corresponding variables must have the same exponents. All constant terms are considered "like" terms.

like terms	unlike terms
$3x$ and $8x$	$9y$ and $10z$
$2x^2y$ and $3x^2y$	$3ab$ and $4ab^2$

In the example, $3x$ and $8x$ are like terms with numerical coefficients of 3 and 8. A **numerical coefficient** of a term is simply the number before its corresponding variable. When combining like terms, simply keep the variable the same and combine the numerical coefficients.

$4y + 10y = 14y$ $6b + 9b - 5b = 10b$
$10x - 3x = 7x$ $12x^2y - 10x^2y + 2x^2y = 4x^2y$

Identify the like terms in each problem.

1. $7c + 12c - 2$
   7c, 12c
2. $19y - 10$
   none
3. $12rt - 10r + 18t$
   none
4. $5r - 10r + 8rs$
   5r, −10r
5. $5d + 7d - 1$
   5d, 7d
6. $q + 9 + 2q + 5q$
   q, 2q, 5q

Simplify. If not possible, write *already simplified*.

7. $8m - 3m$
   5m
8. $8y + 12y + 3y$
   23y
9. $3s + 8s - 2$
   11s − 2
10. $2 + 10k$
    already simplified
11. $8q + 10q + 14$
    18q + 14
12. $4 + 8x + 11y$
    already simplified
13. $5a + 6a - 9a$
    2a
14. $z + 8m + 4z - 4m$
    5z + 4m
15. $5w + 2 + w$
    6w + 2

© Carson-Dellosa • CD-104632

Name _____ 6.EE.A.2b, 6.EE.A.3

Combining Like Terms

$4x + 5y + (-18x) = -14x + 5y$

Combine like terms.

1. $3yz + 5yz =$ **8yz**
2. $3a + 5 + a =$ **4a + 5**
3. $5x - 5y - 8y + 8x =$ **13x − 13y**
4. $18x + 3x =$ **21x**
5. $5 - (-4k) =$ **5 + 4k**
6. $7c - 12c =$ **−5c**
7. $13ab + (-12ab) =$ **ab**
8. $-12x + (-4x) =$ **−16x**
9. $-10n - (-13n) =$ **3n**
10. $12b + (-34b) =$ **−22b**
11. $4.7x - 5.9x =$ **−1.2x**
12. $4x^2 + (-8y) + (-3xy) + 5x^2 + 2xy =$ **$9x^2 - xy - 8y$**
13. $4x + 3y + (-5y) + 3xy + y =$ **4x − y + 3xy**
14. $2x - y + 2x + 3xy =$ **4x + 3xy − y**
15. $5x + 7x =$ **12x**
16. $23x + 8 + 6x + 3y =$ **29x + 3y + 8**
17. $-e + 8e =$ **7e**
18. $2xy + 5x + 6xy + 3xy + (-3x) =$ **2x + 11xy**
19. $7s + 5x - 8s =$ **5x − s**
20. $4xy + 7xy + 6x^2y + 3xy^2 =$ **$11xy + 6x^2y + 3xy^2$**

© Carson-Dellosa • CD-104632

© Carson-Dellosa • CD-104632

Answer Key

Name _____ (6.EE.A.2b, 6.EE.A.3)

Combining Like Terms

Combine like terms.

1. $^-n + 9n + 3 - 8 - 8n =$ **$^-5$**
2. $3(^-4x + 5y) - 3x(2 + 4y) =$ **$^-18x + 15y - 12xy$**
3. $5 - 4y + x + 9y =$ **$5 + x + 5y$**
4. $^-2x + 3y - 5x - {}^-8y + 9y =$ **$^-7x + 20y$**
5. $6(a - b) - 5(2a + 4b) =$ **$^-4a - 26b$**
6. $7(x + 5y) + 3(x + 5y) + 5(3x + 8y) =$ **$25x + 90y$**
7. $12x + 6x + 9x - 3y + (^-7y) + y =$ **$27x - 9y$**
8. $^-21x + (^-2x) =$ **^-23x**
9. $4(x + 9y) - 2(2x + 4y) =$ **$28y$**
10. $4(x + 5y) + (5x + y) =$ **$9x + 21y$**
11. $6x + {}^-2y^2 + 4xy^2 + 3x^2 + 5xy^2 =$ **$3x^2 + 6x - 2y^2 + 9xy^2$**
12. $^-2(c - d) + (c - 3d) - 5(c - d) =$ **$^-6c + 4d$**
13. $3x + (^-3y) - (4x) + y =$ **$^-x - 2y$**
14. $^-3(4x + {}^-2y) - 2(x + 3y) - 2(2x + 6y) =$ **$^-18x - 12y$**
15. $2b + 3(2b + 8a) - 3(8b + 2a) =$ **$18a + {}^-16b$**
16. $3[2(^-y^2 + y) {}^-3] - 3(2x + y) =$ **$^-6y^2 + 3y - 6x - 9$**
17. $2 \cdot 4x \cdot 3y - 4x \cdot 7y =$ **^-4xy**
18. $5(3a^2 - 2b^2) + 3a(a + 3b^2) =$ **$18a^2 + 9ab^2 - 10b^2$**
19. $3c + 4d + 2c + 5d - 4c =$ **$c + 9d$**
20. $4(x^2 + 3y^2) - y(x^2 + 5y) =$ **$4x^2 - xy^2 + 7y^2$**

© Carson-Dellosa • CD-104632

Name _____ (8.EE.C.7b, HSA-REI.B.3)

Solving One-Step Equations (Addition and Subtraction)

When solving equations for a given variable, use the addition or the subtraction property of equality. The addition property means that adding the same number to both sides of an equation will produce an equivalent equation.

$x - 3 = 5$ — Solve for x by adding 3 to both sides of the equation.
$x - 3 + 3 = 5 + 3$
$x = 8$ — When 3 is added to both sides, $^-3$ and 3 gives a result of 0, leaving x by itself. Therefore, $x = 8$ is the answer.

The subtraction property means that subtracting the same number from both sides of an equation will produce an equivalent equation.

$x + 3 = 5$ — Solve for x by subtracting 3 from both sides of the equation.
$x + 3 - 3 = 5 - 3$
$x = 2$ — When 3 is subtracted from both sides, $3 - 3$ gives a result of 0, leaving x by itself. Therefore, $x = 2$ is the answer.

Answers can be checked by substituting the value of x back into the equation. This is to make sure the value of x makes a true sentence when substituted into the original equation.

$x - 3 = 5, x = 8$
$8 - 3 = 5$
$5 = 5$

$x + 3 = 5, x = 2$
$2 + 3 = 5$
$5 = 5$

Solve each equation for x. Check your answers.

1. $x - 10 = 23$ **$x = 33$**
2. $6 + x = {}^-11$ **$x = {}^-17$**
3. $^-13 = {}^-6 + x$ **$x = {}^-7$**
4. $7 = 14 + x$ **$x = {}^-7$**
5. $8 = x + 9$ **$x = {}^-1$**
6. $x - 5 = {}^-5$ **$x = 0$**
7. $x + 3 = 12$ **$x = 9$**
8. $7 + x = 7$ **$x = 0$**
9. $7 + x = {}^-7$ **$x = {}^-14$**
10. $^-2 = x - 5$ **$x = 3$**
11. $x + 4 = {}^-15$ **$x = {}^-19$**
12. $9 = x + 12$ **$x = {}^-3$**

© Carson-Dellosa • CD-104632

Name _____ (8.EE.C.7b, HSA-REI.B.3)

Solving One-Step Equations (Addition and Subtraction)

$12 + x = {}^-24$
$12 - 12 + x = {}^-24 - 12$
$x = {}^-36$

Solve each equation for the given variable.

1. $^-13 + b = 31$ **$b = 44$**
2. $x - 17 = {}^-27$ **$x = {}^-10$**
3. $27 = v + (^-5)$ **$v = 32$**
4. $^-4 = x - 3$ **$x = {}^-1$**
5. $12 - (^-z) = 17$ **$z = 5$**
6. $^-200 = b + (^-73)$ **$b = {}^-127$**
7. $^-13 + x = 18$ **$x = 31$**
8. $^-w + (^-7) = {}^-56$ **$w = 49$**
9. $3 + x = 9$ **$x = 6$**
10. $z + 3.5 = 4.7$ **$z = 1.2$**
11. $12 + (^-g) = 10$ **$g = 2$**
12. $y - 12 = 15$ **$y = 27$**
13. $x + 2 = {}^-6$ **$x = {}^-8$**
14. $s - 5 = {}^-8$ **$s = {}^-3$**
15. $^-13 = n + (^-39)$ **$n = 26$**
16. $r = 4.4 + 3.9$ **$r = 8.3$**

© Carson-Dellosa • CD-104632

Name _____ (8.EE.C.7b, HSA-REI.B.3)

Solving One-Step Equations (Addition and Subtraction)

Solve each equation for the given variable.

1. $c + 23 = 41$ **$c = 18$**
2. $48 - x = 15$ **$x = 33$**
3. $a + 5.7 = 18.9$ **$a = 13.2$**
4. $^-29 - n = 6$ **$n = {}^-35$**
5. $40 - (^-g) = 38$ **$g = {}^-2$**
6. $c - 3 = 4.7$ **$c = 7.7$**
7. $n + \frac{3}{8} = \frac{5}{8}$ **$n = \frac{1}{4}$**
8. $^-1.9 + b = {}^-4.5$ **$b = {}^-2.6$**
9. $z + (^-14) = 13$ **$z = 27$**
10. $^-135 + r = {}^-26$ **$r = 109$**
11. $2\frac{1}{3} + w = 4\frac{2}{9}$ **$w = 1\frac{8}{9}$**
12. $^-0.6 - m = {}^-1.5$ **$m = 0.9$**
13. $^-d + (^-61) = 107$ **$d = {}^-168$**
14. $s - \frac{7}{8} = {}^-\frac{3}{4}$ **$s = \frac{1}{8}$**
15. $^-101 = g - (^-28)$ **$g = {}^-129$**
16. $241 + p = {}^-93$ **$p = {}^-334$**
17. $35.02 - q = 46.1$ **$q = {}^-11.08$**
18. $5\frac{3}{10} + (^-v) = 3.8$ **$v = 1.5$**
19. $^-74 = {}^-k - (^-91)$ **$k = 165$**
20. $^-8 + x = 5.62$ **$x = 13.62$**

© Carson-Dellosa • CD-104632

© Carson-Dellosa • CD-104632

Answer Key

Name _____ 8.EE.C.7b, HSA-REI.B.3

Solving One-Step Equations (Multiplication and Division)

When solving equations for a given variable, use the multiplication or division property of equality. The multiplication property means that multiplying the same nonzero number by both sides of the equation will produce an equivalent equation.

$\frac{x}{7} = 3$ Multiply both sides of the equation by 7 to isolate x.

$(7)(\frac{x}{7}) = (7)(3)$ Check $\frac{21}{7} = 3$

$x = 21$ $3 = 3$

Therefore, the solution is 21.

The division property means that dividing both sides of the equation by the same nonzero number will produce an equivalent equation.

$-5x = 20$ Divide both sides of the equation by -5 to isolate x.

$\frac{-5x}{-5} = \frac{20}{-5}$ Check $-5(-4) = 20$

$x = -4$ $20 = 20$

Therefore, the solution is -4.

Solve each equation for x. Check your answers.

1. $5x = 35$ **$x = 7$**
2. $18 = -3x$ **$x = -6$**
3. $-7x = 49$ **$x = -7$**
4. $-\frac{1}{3}x = 6$ **$x = -18$**
5. $-5x = -20$ **$x = 4$**
6. $-\frac{5}{8}x = 10$ **$x = -16$**
7. $\frac{1}{4}x = -2$ **$x = -8$**
8. $\frac{2}{3}x = 8$ **$x = 12$**
9. $4 = -\frac{x}{5}$ **$x = -20$**
10. $-4x = 48$ **$x = -12$**
11. $\frac{x}{3} = -5$ **$x = -15$**
12. $-6x = 24$ **$x = -4$**

© Carson-Dellosa • CD-104632

Name _____ 8.EE.C.7b, HSA-REI.B.3

Solving One-Step Equations (Multiplication and Division)

$3x = 15$

$\frac{3x}{3} = \frac{15}{3}$

$x = 5$

$-\frac{3}{4y} = -6$

$-\frac{4}{3} \cdot -\frac{3}{4y} = -6 \cdot -\frac{4}{3}$

$\frac{1}{y} = 8$ so $y = \frac{1}{8}$

Solve each equation for the given variable.

1. $-13h = 169$ **$h = -13$**
2. $4b = -36$ **$b = -9$**
3. $10x = -100$ **$x = -10$**
4. $4c = 288$ **$c = 72$**
5. $7x = -63$ **$x = -9$**
6. $4y = -48$ **$y = -12$**
7. $6x = -36$ **$x = -6$**
8. $\frac{8}{k} = \frac{2}{5}$ **$k = 20$**
9. $-(-90) = -45z$ **$z = -2$**
10. $-50 = 2x$ **$x = -25$**
11. $\frac{2}{n} = \frac{1}{9}$ **$n = 18$**
12. $\frac{4}{x} = \frac{2}{9}$ **$x = 18$**
13. $\frac{x}{6} = \frac{6}{9}$ **$x = 4$**
14. $-35c = 700$ **$c = -20$**
15. $-4x = -20$ **$x = 5$**
16. $-\frac{x}{6} = \frac{2}{3}$ **$x = -4$**

© Carson-Dellosa • CD-104632

Name _____ 8.EE.C.7b, HSA-REI.B.3

Solving One-Step Equations (Multiplication and Division)

Solve each equation for the given variable.

1. $12.8 = 4b$ **$b = 3.2$**
2. $-\frac{3}{4} = \frac{n}{16}$ **$n = -12$**
3. $\frac{2}{3}x = -10$ **$x = -15$**
4. $-(-36) = -0.25c$ **$c = -144$**
5. $1.6r = 80$ **$r = 50$**
6. $-\frac{x}{8} = \frac{1}{4}$ **$x = -2$**
7. $208x = -4$ **$x = -\frac{1}{52}$**
8. $\frac{2}{k} = \frac{1}{8}$ **$k = 16$**
9. $-4.9 = 7m$ **$m = -0.7$**
10. $0.006w = -0.54$ **$w = -90$**
11. $\frac{720}{p} = 9$ **$p = 80$**
12. $-\frac{5h}{50} = -0.1$ **$h = 1$**
13. $-9s = 6$ **$s = -\frac{2}{3}$**
14. $15d = -\frac{1}{3}$ **$d = -\frac{1}{45}$**
15. $-(-27)d = 16.2$ **$d = 0.6$**
16. $3z = -174$ **$z = -58$**
17. $-\frac{4}{5j} = -8$ **$j = \frac{1}{10}$**
18. $\frac{56}{63} = -\frac{x}{9}$ **$x = -8$**

© Carson-Dellosa • CD-104632

Name _____ 8.EE.C.7b, HSA-REI.B.3

Solving Basic Multistep Equations

When solving equations for a given variable, sometimes you need to use more than one of the properties of equality.

$3x - 2 = 7$

$3x - 2 + 2 = 7 + 2$ Add 2 to both sides of the equation.

$\frac{3x}{3} = \frac{9}{3}$ Now, divide both sides of the equation by 3.

$x = 3$ Check $3(3) - 2 = 7$

$9 - 2 = 7$

$7 = 7$

Therefore, the solution is 3.

Here are some steps to follow when solving multi-step equations.

1. Simplify both sides of the equation (if possible).
2. Use the addition or subtraction property of equality to isolate terms containing the variable.
3. Use the multiplication or the division property of equality to further isolate the variable.
4. Check the solution.

Solve each equation for x. Check your answers.

1. $6x - 3 = 21$ **$x = 4$**
2. $-6 + \frac{x}{4} = 1$ **$x = 28$**
3. $18 - 3x = -12$ **$x = 10$**
4. $7 + 2x = -13$ **$x = -10$**
5. $-4 = 7x + 8 - 8x$ **$x = 12$**
6. $13 = 9 - \frac{x}{5}$ **$x = -20$**
7. $-7 - x = -5$ **$x = -2$**
8. $5x + 9 - 4x = 12$ **$x = 3$**
9. $-8x - 13 = 19$ **$x = -4$**
10. $-3 = -5 - 2x$ **$x = -1$**
11. $\frac{1}{2}x + 9 = 15$ **$x = 12$**
12. $-7 = 3x - 15 - 7x$ **$x = -2$**

© Carson-Dellosa • CD-104632

© Carson-Dellosa • CD-104632

Answer Key

Name _____ 8.EE.C.7b, HSA-REI.B.3

Solving Basic Multistep Equations

$$4x + 4 = 12$$
$$4x + 4 - 4 = 12 - 4$$
$$4x = 8$$
$$x = 2$$

Solve each equation for the given variable.

1. $7x - 12 = 2$
   $x = 2$
2. $7a - 4 = 24$
   $a = 4$
3. $4b - 7 = 37$
   $b = 11$
4. $3c - 9 = 9$
   $c = 6$
5. $8 - 9y = 35$
   $y = -3$
6. $8 - 12x = 32$
   $x = -2$
7. $1.3x + 5 = -5.4$
   $x = -8$
8. $3(y + 4) + 5 = 35$
   $y = 6$
9. $0 = 25x + 75$
   $x = -3$
10. $3 - \frac{1}{5}e = -7$
    $e = 50$
11. $5 - \frac{1}{2}x = -9$
    $x = 28$
12. $2x = 6 + (-18)$
    $x = -6$
13. $7 - \frac{1}{9}k = 32$
    $k = -225$
14. $\frac{3}{12}w + 2 = 11$
    $w = 36$
15. $\frac{2x}{5} + 3 = 9$
    $x = 15$
16. $\frac{x}{3} - 8 = -12$
    $x = -12$
17. $5(e + 5) = -10$
    $e = -7$
18. $8 - \frac{1}{2}y = -6$
    $y = 28$

Name _____ 8.EE.C.7b, HSA-REI.B.3

Solving Basic Multistep Equations

Solve each equation for the given variable.

1. $5z - 8 = -28$
   $z = -4$
2. $4k + 7 = -9$
   $k = -4$
3. $13x + 7 = -32$
   $x = -3$
4. $2x + 12 = 6$
   $x = -3$
5. $7.2 + 4x = 19.2$
   $x = 3$
6. $2(w - 6) = 8$
   $w = 10$
7. $7h + 1 = -13$
   $h = -2$
8. $3(c - 2) = 15$
   $c = 7$
9. $6x - 5 = -41$
   $x = -6$
10. $-3 + 2n = -15$
    $n = -6$
11. $5e + (-9) = 26$
    $e = 7$
12. $\frac{m}{3} - 7 = -10$
    $m = -9$
13. $6x - 2 = 34$
    $x = 6$
14. $-8(r - 2) = 40$
    $r = -3$
15. $5n - 8 = -23$
    $n = -3$
16. $2 + (\frac{1}{5})x = -7$
    $x = -45$
17. $5 - (\frac{1}{2})g = 12$
    $g = -14$
18. $3x - 4 = 14$
    $x = 6$
19. $-6 = \frac{3z}{4} + 12$
    $z = -24$
20. $2(f + 7) - 8 = 22$
    $f = 8$
21. $4.7 = -3.4m - 5.5$
    $m = -3$
22. $32 = \frac{4}{6}x - 34$
    $x = 99$

Name _____ 8.EE.C.7b, HSA-REI.B.3

Solving Equations with Variables on Both Sides

Often, equations with a variable on both sides of the equation need to be solved. This requires one additional step of getting the variable on one side only.

$4x - 1 = 2x + 7$

$4x - 2x - 1 = 2x - 2x + 7$ Get the variable on one side of the equation.

$2x - 1 = 7$ Now, solve for x using the properties of equality.

$2x - 1 + 1 = 7 + 1$

$\frac{2x}{2} = \frac{8}{2}$

$x = 4$

Check $4(4) - 1 = 2(4) + 7$

$16 - 1 = 8 + 7$

$15 = 15$

Therefore, the solution is 4.

Combine the variables on the side of the equation with the greater variable coefficient, in order to avoid solving an equation with a variable which has a negative coefficient.

Solve each equation for x. Check your answers.

1. $9x - 12 = 3x$
   $x = 2$
2. $8x - 12 = 15x - 4x$
   $x = -4$
3. $11 + 6x = 2x - 13$
   $x = -6$
4. $-5x = 9 - 2x$
   $x = -3$
5. $-8x - 10 = 4x + 14$
   $x = -2$
6. $10x - 5 = 21 - 3x$
   $x = 2$
7. $-12x = 14 - 5x$
   $x = -2$
8. $19 - 3x = 21 + x$
   $x = -\frac{1}{2}$
9. $4x + 12 = -3x - 6 + 4x$
   $x = -6$
10. $14x + 5 = 8x - 1$
    $x = -1$
11. $7x - 3 = -4x - 25$
    $x = -2$
12. $-9x + 5 = -22 - 6x$
    $x = 9$

Name _____ 8.EE.C.7b, HSA-REI.B.3

Solving Equations with Variables on Both Sides

$$6x - 7 = x + 33$$
$$6x - x - 7 = x - x + 33$$
$$5x - 7 + 7 = 33 + 7$$
$$5x = 40$$
$$x = 8$$

Solve each equation for the given variable.

1. $4x - 6 = 8 - 3x$
   $x = 2$
2. $2(x + 3) = 12 - x$
   $x = 2$
3. $5(5 - z) = 4(z - 5)$
   $z = 5$
4. $4e + 6 = -8 + 11e$
   $e = 2$
5. $-4(x - 6) = 2(7 - 7x)$
   $x = -1$
6. $2m - 9 = 8m - 27$
   $m = 3$
7. $b + 9 = 6 + 2b$
   $b = 3$
8. $-9j + 3 = -32 - 4j$
   $j = 7$
9. $5(j - 4) = -8 - j$
   $j = 2$
10. $8(-9x + 4) = -3(6x + 9) + 5$
    $x = 1$
11. $3d - 5 = -9 + 2d$
    $d = -4$
12. $-2 + 6d = 34 - 3d$
    $d = 4$
13. $3(k + 4) = -3 - 2k$
    $k = -3$
14. $5(m - 3) = 27 - 2m$
    $m = 6$
15. $-j + 2 = -14 - 5j$
    $j = -4$
16. $-(x + 7) - 5 = 4(x + 3) - 6x$
    $x = 24$
17. $5(x - 1) = 2x + 4(x - 1)$
    $x = -1$
18. $-3(5k + 5) = 54 - 12k$
    $k = 23$
19. $3e + 1 = 36 - 4e$
    $e = 5$
20. $-6r + 12 = -2 + 8r$
    $r = 1$

Answer Key

Name _____ (8.EE.C.7b, HSA-REI.B.3)

Solving Equations with Variables on Both Sides

Solve each equation for the given variable.

1. $7 - 6a = 6 - 7a$
   $a = -1$
2. $3c - 12 = 14 + 5c$
   $c = -13$
3. $3x - 3 = -3x + -3$
   $x = 0$
4. $2x - 7 = 3x + 4$
   $x = -11$
5. $9a + 5 = 3a - 1$
   $a = -1$
6. $8(x - 3) + 8 = 5x - 22$
   $x = -2$
7. $5d + 7 = 4d - 9$
   $d = -16$
8. $-10w + 6 = -7w + -9$
   $w = 5$
9. $-7c + 9 = c + 1$
   $c = 1$
10. $2n + 6 = 5n - 9$
    $n = 5$
11. $\frac{5}{2}x + 3 = \frac{1}{2}x + 15$
    $x = 6$
12. $5 + 3x = 7(x + 3)$
    $x = -4$
13. $12m - 9 = 4m + 15$
    $m = 3$
14. $2(f - 4) + 8 = 3f - 8$
    $f = 8$
15. $-6 - (-2n) = 3n - 6 + 5$
    $n = -5$
16. $4(2y - 4) = 5y + 2$
    $y = 6$
17. $2(r - 4) = 5[r + (-7)]$
    $r = 9$
18. $6(x - 9) = 4(x - 5)$
    $x = 17$
19. $4(z + 5) - 3 = 6z - 13$
    $z = 15$
20. $4e - 19 = -3(e + 4)$
    $e = 1$
21. $4(-6x - 2) = -22x$
    $x = -4$
22. $-\frac{1}{3}x - 5 = 7 - x$
    $x = 18$

© Carson-Dellosa • CD-104632

Name _____ (6.EE.A.2, 7.EE.B.3, 8.EE.C.7b)

Writing and Solving Equations

The sum of three times a number and 25 is 40. Find the number.

$3x + 25 = 40$
$3x + 25 - 25 = 40 - 25$
$3x = 15$
$x = 5$ The number is 5.

Write an equation for each word problem and solve it.

1. The difference of a number and −3 is 8. Find the number.
   Equation **$x - (-3) = 8$** **Solution** **$x = 5$**
2. Twice a number added to 9 is 15. Find the number.
   Equation **$2x + 9 = 15$** **Solution** **$x = 3$**
3. Twelve subtracted from 3 times a number is 15. Find the number.
   Equation **$3x - 12 = 15$** **Solution** **$x = 9$**
4. The sum of 4 times a number and 5 is −7. Find the number.
   Equation **$4x + 5 = -7$** **Solution** **$x = -3$**
5. The product of a number and 5 is 60. Find the number.
   Equation **$5x = 60$** **Solution** **$x = 12$**
6. The difference of 5 times a number and 6 is 14. Find the number.
   Equation **$5x - 6 = 14$** **Solution** **$x = 4$**
7. The sum of a number and −6 is 10. Find the number.
   Equation **$x + (-6) = 10$** **Solution** **$x = 16$**
8. The quotient of a number and 4 is −12. Find the number.
   Equation **$\frac{x}{4} = -12$** **Solution** **$x = -48$**

© Carson-Dellosa • CD-104632

Name _____ (6.EE.A.2, 7.EE.B.3, 8.EE.C.7b)

Writing and Solving Equations

Write an equation for each word problem and solve it.

1. Six times the difference of a number and 9 is 54. Find the number.
   Equation **$6(x - 9) = 54$** **Solution** **$x = 18$**
2. The sum of 8 times a number and 3 is 59. Find the number.
   Equation **$8x + 3 = 59$** **Solution** **$x = 7$**
3. The sum of 5 times a number and −11 is −16. Find the number.
   Equation **$5x + (-11) = -16$** **Solution** **$x = -1$**
4. Twelve times the sum of a number and −8 is 48. Find the number.
   Equation **$12[x + (-8)] = 48$** **Solution** **$x = 12$**
5. The sum of 5 times a number and 2 is −13. Find the number.
   Equation **$5x + 2 = -13$** **Solution** **$x = -3$**
6. The sum of 7 times a number and 11 is 81. Find the number.
   Equation **$7x + 11 = 81$** **Solution** **$x = 10$**
7. Three times the sum of a number and −2 is −15. Find the number.
   Equation **$3[x + (-2)] = -15$** **Solution** **$x = -3$**
8. Five times the sum of a number and 2 is 35. Find the number.
   Equation **$5(x + 2) = 35$** **Solution** **$x = 5$**

© Carson-Dellosa • CD-104632

Name _____ (6.EE.A.2, 7.EE.B.3, 8.EE.C.7b)

Writing and Solving Equations

Write an equation for each word problem and solve it.

1. A number increased by 10 is 38. Find the number.
   Equation **$x + 10 = 38$** **Solution** **$x = 28$**
2. A number times 8 is −96. Find the number.
   Equation **$8x = -96$** **Solution** **$x = -12$**
3. Thirty-two is 7 less than 3 times a number. Find the number.
   Equation **$32 = 3x - 7$** **Solution** **$x = 13$**
4. A number multiplied by 4 and increased by 7 is −25. Find the number.
   Equation **$-25 = 4x + 7$** **Solution** **$x = -8$**
5. A number decreased by 12 is the same as 3 times the number. Find the number.
   Equation **$x - 12 = 3x$** **Solution** **$x = -6$**
6. Four times the sum of a number and 7 is 44 less than the number. Find the number.
   Equation **$4(x + 7) = x - 44$** **Solution** **$x = -24$**
7. A number increased by 16 is the same as 8 times the sum of the number and 9. Find the number.
   Equation **$16 + x = 8(x + 9)$** **Solution** **$x = -8$**
8. Four times a number is decreased by 9 and then increased by 12. The result is 5 less than 2 times the number. Find the number.
   Equation **$4x - 9 + 12 = 2x - 5$** **Solution** **$x = -4$**

© Carson-Dellosa • CD-104632

© Carson-Dellosa • CD-104632

Answer Key

Name _____ 6.EE.B.8, HSA-REI.B.3

Solving Inequalities

$$-10n + 5 \le 55$$
$$-10n + 5 - 5 \le 55 - 5$$
$$-10n \le 50$$
$$n \ge -5$$

Solve each inequality and graph its solution set.

1. $-4(3d + 2) \le 4$
   $d \ge -1$
2. $10 - 5x - 20 \ge -20$
   $x \le 2$
3. $-15 > 4y - 7 - 3y - 4$
   $y < -4$
4. $4x - 7 < 9$
   $x < 4$
5. $6n - 3 > 33$
   $n > 6$
6. $3(3c - 4) \le 15$
   $c \le 3$
7. $5z - 1 > 9$
   $z > 2$
8. $5 > 4x - 11$
   $x < 4$

© Carson-Dellosa • CD-104632 41

Name _____ 6.EE.B.8, HSA-REI.B.3

Solving Inequalities

$$-10x > 4x - 42$$
$$-10x + 10x > 4x + 10x - 42$$
$$0 > 14x - 42$$
$$0 + 42 > 14x - 42 + 42$$
$$42 > 14x$$
$$x < 3$$

Solve each inequality and graph its solution set.

1. $-2a < 10 + 3a$
   $a > -2$
2. $4w > 2w + 8$
   $w > 4$
3. $2(k + 4) \le 3(2k - 4)$
   $k \ge 5$
4. $5n + 3 \ge -12$
   $n \ge -3$
5. $5c + 2 < 2c + (-7)$
   $c < -3$
6. $5x - 20 > 2x + 1$
   $x > 7$
7. $3(s - 4) \ge 4s - 12$
   $s \le 0$
8. $-9 - e > 3e + 11$
   $e < -5$

42 © Carson-Dellosa • CD-104632

Name _____ 6.EE.B.8, HSA-REI.B.3

Solving Inequalities

Solve each inequality and graph its solution set.

1. $15e - 3 \le 20e + 17$
   $e \ge -4$
2. $6x - 4 > 2(x - 6)$
   $x > -2$
3. $7c - 8 \ge 6$
   $c \ge 2$
4. $5x + (-3) \ge 2(3 + x)$
   $x \ge 3$
5. $-8 < 2(2 + 3r)$
   $r > -2$
6. $6d < 3d - 18$
   $d < -6$
7. $7m + 9 \le 6(m + 3)$
   $m \le 9$
8. $3(2x + 4) \ge 7x + 8$
   $x \le 4$
9. $-4.2 > 0.6y$
   $y < -7$
10. $\frac{a}{4} + 3 \le 5$
    $a \le 8$

© Carson-Dellosa • CD-104632 43

Name _____ 8.EE.A.1

Multiplication and Powers of Exponents

When multiplying exponents, it is important to remember the following properties:

1. When multiplying powers having the same base, add the exponents, keeping the same base. (Remember: in a^3, a is the base, 3 is the exponent, and a^3 is the power.) For example, $x^3 \bullet x^5 = x^{3+5} = x^8$.
2. When finding a power of a power, multiply the exponents. For example, $(x^3)^2 = x^6$.
3. When finding the power of a product, find the power of each factor and multiply. For example, $(x \bullet y)^2 = x^2 \bullet y^2$.

Simplify $(5x^3)^2(xy)^3$
$(5^2x^6)(x^3y^3)$
$= 25x^9y^3$

Simplify $(2x^4)^3(-x^2)^3$
$(2^3x^{12})(-x^6)$
$= -8x^{18}$

1. What do you do with the exponents when multiplying powers that have the same base?
   Add the exponents.
2. Label the base, the exponent, and the power in x^3.
   x: base, 3: exponent, x^3: power
3. Explain what you are to do when finding the power of a product.
   Find the power of each factor and multiply.

Simplify each expression.

4. $3x \bullet x^2$
   $3x^3$
5. $(-6xy)^2(x^2y)^3$
   $36x^8y^5$
6. $(8xy)^2$
   $64x^2y^2$
7. $-4x^4 \bullet x^3$
   $-4x^7$
8. $(-3x^2y)^3$
   $-27x^6y^3$
9. $(4c^2)(-5c^7)$
   $-20c^9$
10. $(-x^2)(-x)^2$
    x^4
11. $-xy(-xy)^2$
    $-x^3y^3$
12. $(3x^3)(5x^5)$
    $15x^8$
13. $(x^3y^3)^3$
    x^9y^9

44 © Carson-Dellosa • CD-104632

Answer Key

Name _____ 8.EE.A.1

Multiplication and Powers of Exponents

Rule: $(x^a)^b = x^{ab}$ Example: $(x^2y^3)^3 = x^6y^9$

Rule: $x^a \bullet x^b = x^{a+b}$ Example: $x^3 \bullet x^5 = x^8$

Simplify each expression.

1. $c \bullet c^2 \bullet c^3 =$ **c^6**
2. $e \bullet e^2 \bullet e^3 \bullet e^4 \bullet e^5 =$ **e^{15}**
3. $a^3 \bullet a^4 \bullet a^7 \bullet a =$ **a^{15}**
4. $(3xy^2)(2x^2y^3) =$ **$6x^3y^5$**
5. $(2a^2b)(4ab^2) =$ **$8a^3b^3$**
6. $(5f)(-3f^3)(2f)^2 =$ **$-60f^6$**
7. $(m^2n)^3(4mn^2)(mn) =$ **$4m^8n^6$**
8. $(4k^2)(-3k)(3k^5) =$ **$-36k^8$**
9. $(-2c^4)(4cd)(-cd^2) =$ **$8c^6d^3$**
10. $(3x^3)^2(3x^4)(-3x^2)^3 =$ **$-729x^{16}$**
11. $(-1)(x)(-x^2)(x^3)(-x^2) =$ **$-x^8$**
12. $(3x^2)(-3x^5) =$ **$-9x^7$**
13. $(c^2h)^2(3ch^3)(2c^3h^4)^2 =$ **$12c^{11}h^{13}$**
14. $(-4p^3)(-4p^6)(-2p^9) =$ **$-32p^{18}$**
15. $(12c^3)(2g^3)(4ch)^3$ **$1{,}536c^6g^3h^3$**
16. $(4x^2y^3)^3(x^3y)(-x^2y^2) =$ **$-64x^{11}y^{12}$**
17. $(-4f^3)^4(-3m^3)^5 =$ **$62{,}208f^{12}m^{15}$**
18. $(2c^2d^2)^4(-5cd^4) =$ **$-80c^9d^{12}$**
19. $(3x)^2(2x^3y^4)(-5x^6y^2) =$ **$-90x^{11}y^8$**
20. $(3x)(-4y^2)(6x^3y) =$ **$-72x^4y^3$**

Name _____ 8.EE.A.1

Multiplication and Powers of Exponents

Simplify each expression.

1. $(-4xy^3)^3 =$ **$-64x^3y^9$**
2. $(x^2y^3)(x^3y) =$ **x^5y^4**
3. $(-6x^4y^6)^3 =$ **$-216x^{12}y^{18}$**
4. $(5x^2y^4)^3 =$ **$125x^6y^{12}$**
5. $(6x^5y^4)(5x^5y^4) =$ **$30x^{10}y^8$**
6. $(2x)^4(x^3y)^2 =$ **$16x^{10}y^2$**
7. $(4x^3y^2)^3(-2x^2y^4) =$ **$-128x^{11}y^{10}$**
8. $(3x^2y)(-8xy^4) =$ **$-24x^3y^5$**
9. $(-2x^2y)^4 =$ **$16x^8y^4$**
10. $(x^2y^3)^2(x^3y^2)^4 =$ **$x^{16}y^{14}$**
11. $(-4x^3y^3)^4(-8x^3)^2 =$ **$16{,}384x^{18}y^{12}$**
12. $(3xy^3)(-4x^2y^4)^2(xy^3) =$ **$48x^6y^{14}$**
13. $(-3x^3y)^3 =$ **$-27x^9y^3$**
14. $(-2x^4y^5)^3 =$ **$-8x^{12}y^{15}$**
15. $(3x^2y^3)^4(x^3y)(7xy^3) =$ **$567x^{12}y^{16}$**
16. $(6x^2y^3)(4xy^2)^3(3x^2y)^2 =$ **$3{,}456x^9y^{11}$**
17. $(-2x^3y^3z)^4(2xyz^4)^2 =$ **$64x^{14}y^{14}z^{12}$**
18. $(5xy^3)(-5xy^2)^5 =$ **$-15{,}625x^6y^{13}$**
19. $(-3x^2y^3)^2 =$ **$9x^4y^6$**
20. $(4z^4)^2(2x^2y)(-3xy^3z^5) =$ **$-96x^3y^4z^{13}$**

Name _____ 8.EE.A.1

Division of Exponents

When dividing exponents, it is important to remember the following properties:

1. When dividing powers that have the same base, subtract the exponents.

   For example, $\frac{x^4}{x^2} = x^{4-2} = x^2$, where x cannot be equal to 0.

2. When finding a power of a quotient, find the power of the numerator and the power of the denominator and divide.

   For example, $(\frac{x}{y})^3 = \frac{x^3}{y^3}$, where y cannot be equal to 0.

   Simplify: $\frac{6^8}{6^6} = 6^{8-6}$, $6^2 = 36$ $\quad$ $(\frac{3}{4})^{-2} = \frac{3^{-2}}{4^{-2}} = \frac{4^2}{3^2} = \frac{16}{9}$

1. Explain what you do with the exponents when dividing powers that have the same base.

   Subtract the exponents.

Evaluate each expression.

2. $\frac{5^6}{5^3}$ **125**
3. $\frac{(-3)^2}{3^2}$ **1**
4. $\frac{3^5}{3^2}$ **27**
5. $\frac{5^4 \bullet 5}{5^7}$ **$\frac{1}{25}$**
6. $\frac{8^6}{8^3}$ **512**
7. $\frac{7^3}{7}$ **49**
8. $\frac{4^8}{4^8}$ **1**
9. $\frac{6^4 \bullet 6^3}{6^5}$ **36**
10. $\frac{4^6}{4^2}$ **256**
11. $\frac{10^5}{10^3}$ **100**

Simplify each expression.

12. $\frac{x^5}{x^3}$ **x^2**
13. $x^7 \bullet \frac{1}{x^5}$ **x^2**
14. $\frac{18x^4y^4}{6x^2y^4}$ **$3x^2$**
15. $\frac{4x^4y^4}{4x^2y}$ **x^2y^3**
16. $\frac{6x^2y^4}{3y^2}$ **$2x^2y^2$**
17. $x^4 \bullet \frac{1}{x^2}$ **x^2**

Name _____ 8.EE.A.1

Division of Exponents

Rule: $\frac{x^a}{x^b} = x^{a-b}$ Example: $\frac{x^6}{x^4} = x^{6-4} = x^2$ $\quad$ $\frac{x^3}{x^{-2}} = x^{3-(-2)} = x^5$

Simplify each expression.

1. $\frac{-12m^5}{6m} =$ **$-2m^4$**
2. $\frac{x^3}{x^2} =$ **x**
3. $\frac{9a^3b^5}{-3ab^2} =$ **$-3a^2b^3$**
4. $\frac{16c^3}{-4c^2} =$ **$-4c$**
5. $\frac{d^3}{d^2} =$ **d**
6. $\frac{-3p^8}{6p^2} =$ **$\frac{(-p^6)}{2}$**
7. $\frac{-54c^2d^4}{-8cd} =$ **$\frac{27cd^3}{4}$**
8. $\frac{49r^{13}}{-7r^8} =$ **$-7r^5$**
9. $\frac{45k^7r^3}{-3k^5} =$ **$-15k^2r^3$**
10. $\frac{-14c^{15}d^3}{-2c^9d} =$ **$7c^6d^2$**
11. $\frac{(5k)(-8k^5)}{10k^3} =$ **$-4k^3$**
12. $\frac{24x^2y}{-4x^2} =$ **$-6y$**
13. $\frac{4x^2y^3z^4}{2xy^2z^3} =$ **$2xyz$**
14. $\frac{9a^{11}}{a^3} =$ **$9a^8$**
15. $\frac{22y^5z^8}{2yz^7} =$ **$11y^4z$**
16. $\frac{b^{14}c^6}{b^5c^4} =$ **b^9c^5**

© Carson-Dellosa • CD-104632

Answer Key

Name _____ 8.EE.A.1

Division of Exponents

Simplify.

1. $\frac{a^5}{a^3} =$ **a^2**
2. $\frac{a^5b^2}{2a^2} =$ **$\frac{a^3b^2}{2}$**
3. $\frac{13m^6n^7}{39m^3n^5} =$ **$\frac{m^3n^2}{3}$**
4. $\frac{9x^8y^7z^8}{18x^5y^5z^4} =$ **$\frac{x^3y^2z^4}{2}$**
5. $\frac{-2c^2d^3}{-8cd^2} =$ **$\frac{cd}{4}$**
6. $\frac{10a^6b^8}{40a^2b^2} =$ **$\frac{a^4b^6}{4}$**
7. $\frac{18a^6b^2c^6}{36a^4bc^2} =$ **$\frac{a^2bc^4}{2}$**
8. $\frac{5x^3y^2z^2}{5x^2yz} =$ **xyz**
9. $\frac{45x^9y^{10}z^5}{51x^9y^8z^3} =$ **$\frac{15y^2z^2}{17}$**
10. $\frac{16x^2y^4}{4x^2y^3} =$ **$4y$**
11. $\frac{18x^6y^3z^4}{12x^3y^2z^3} =$ **$\frac{3x^3yz}{2}$**
12. $\frac{72x^5y^5z^6}{8x^4yz^3} =$ **$9xy^4z^3$**
13. $\frac{18a^9b^3}{12a^2b^2} =$ **$\frac{3a^7b}{2}$**
14. $\frac{44x^8y^2}{11x^7y} =$ **$4xy$**
15. $\frac{(6x^3)(3x^8)}{-12x^{10}} =$ **$\frac{-3x}{2}$**
16. $\frac{21k^9}{(3k)(7k^4)} =$ **k^4**
17. $\frac{(110c^3)(-c^9)}{11c^5} =$ **$-10c^7$**
18. $\frac{(3xy)(4x^2y)}{-6xy^2} =$ **$-2x^2$**

Name _____ 8.EE.A.1

Negative and Zero Exponents

Given a nonzero number a and a positive integer n, the following definitions of negative exponents and zero exponents are stated.

1. For a negative exponent: the expression a^{-n} is the reciprocal of a^n. This is written: $a^{-n} = \frac{1}{a^n}$, where $a \neq 0$.
2. For a zero exponent: any nonzero number raised to the 0 power will have an answer of 1. This is written: $a^0 = 1$, where $a \neq 0$.

Evaluate: 5^{-3} $= \frac{1}{5^3}$ $= \frac{1}{125}$

Simplify by rewriting with positive exponents. $2x^{-3}y^2z^{-4}$ $= 2 \cdot \frac{1}{x^3} \cdot y^2 \cdot \frac{1}{z^4}$ $= \frac{2y^2}{x^3z^4}$

Note: When rewriting expressions to positive exponent form, simply move the factors from the denominator to the numerator or vice versa, leaving out the in-between step.

1. Any number raised to the zero power has what value?

   1

Evaluate each expression.

2. 4^{-2} **$\frac{1}{16}$**
3. $7^3 \cdot 7^{-3}$ **1**
4. $3 \cdot 3^{-1}$ **1**
5. $-5^0 \cdot \frac{1}{3^{-3}}$ **-27**
6. $(6^2)^{-2}$ **$\frac{1}{1296}$**
7. $(-2^{-3})^{-1}$ **-8**

Rewrite each expression with positive exponents.

8. x^{-8} **$\frac{1}{x^8}$**
9. $\frac{1}{3x^{-2}}$ **$\frac{x^2}{3}$**
10. $(-3)^0x^{-2}$ **$\frac{1}{x^2}$**
11. x^{-10} **$\frac{1}{x^{10}}$**
12. $x^{-3}y^4$ **$\frac{y^4}{x^3}$**
13. $\frac{6}{x^{-2}}$ **$6x^2$**

Name _____ 8.EE.A.1

Negative and Zero Exponents

Rule: $x^{-a} = \frac{1}{x^a}$ Example: $4^{-2} = \frac{1}{16}$ Example: $4x^{-2} = \frac{4}{x^2}$ Example: $(2x)^{-3} = \frac{1}{8x^3}$

$4^{-2} = \frac{1}{4^2} = \frac{1}{16}$ $\frac{1}{(2x)^3} = \frac{1}{8x^3}$

Rule: $x^0 = 1$

Simplify.

1. $4cd^{-5}$ **$\frac{4c}{d^5}$**
2. 3^0 **1**
3. $3a^4b^{-3}$ **$\frac{3a^4}{b^3}$**
4. 4^{-5} **$\frac{1}{1024}$**
5. $(-2)^{-2}$ **$\frac{1}{4}$**
6. $(3xy)^{-1}$ **$\frac{1}{3xy}$**
7. $(3x)^{-3}$ **$\frac{1}{27x^3}$**
8. $7x^{-3}$ **$\frac{7}{x^3}$**
9. $-2x^{-3}$ **$\frac{-2}{x^3}$**
10. $(6y^2)^{-2}$ **$\frac{1}{36y^4}$**
11. $(\frac{4}{5})^{-2}$ **$\frac{25}{16}$**
12. $4m^3n^{-5}$ **$\frac{4m^3}{n^5}$**
13. $(-11x^3y)^{-2}$ **$\frac{1}{121x^6y^2}$**
14. $(c^0d)^{-2}$ **$\frac{1}{d^2}$**
15. $(\frac{x^2}{y^3})^{-2}$ **$\frac{y^6}{x^4}$**
16. $(\frac{2}{3})^{-1}$ **$\frac{3}{2}$**
17. b^0 **1**
18. c^{-7} **$\frac{1}{c^7}$**

Name _____ 8.EE.A.1

Negative and Zero Exponents

Simplify.

1. $6x^{-3}$ **$\frac{6}{x^3}$**
2. $4x^{-5}$ **$\frac{4}{x^5}$**
3. $\frac{1}{5x^{-4}}$ **$\frac{x^4}{5}$**
4. $4x^{-4}y^{-2}$ **$\frac{4}{x^4y^2}$**
5. $\frac{1}{(2x)^{-3}}$ **$8x^3$**
6. $(3x^{-2})^2$ **$\frac{9}{x^4}$**
7. $14x^{-8}y$ **$\frac{14y}{x^8}$**
8. $(-5x^3)^{-2}$ **$\frac{1}{25x^6}$**
9. $(-3x^{-2}y^5z^3)^{-2}$ **$\frac{x^4}{9y^{10}z^6}$**
10. $8x^{-2}y^{-4}$ **$\frac{8}{x^2y^4}$**
11. $2x^6y^{-5}z^2$ **$\frac{2z^2}{y^5}$**
12. $(4x^3y^{-4})^2$ **$\frac{16x^6}{y^8}$**
13. $(\frac{x^3}{2y^{-2}})^{-3}$ **$\frac{8}{x^9y^6}$**
14. $4m^{-5}n^{-1}$ **$\frac{4}{m^5n}$**
15. $(\frac{x^2}{y^3})^{-2}$ **$\frac{y^6}{x^4}$**
16. $(-14c^2d)^{-1}$ **$\frac{1}{-14c^2d}$**
17. $9xy^{-4}$ **$\frac{9x}{y^4}$**
18. $(-6s^3p^2)^{-3}$ **$\frac{1}{-216s^9p^6}$**

Answer Key

Name _____ (7.EE.A.1, HSA-APR.A.1)

Adding and Subtracting Polynomials

When adding polynomials, the following two methods can be used.

1. The horizontal method:

   Add $(3x^4 + 8x^2 - 2)$ and $(^-4x^2 + 7x^4 + 5)$.
   First group like terms in descending order. $(3x^4 + 7x^4) + (8x^2 - 4x^2) + (^-2 + 5)$
   Simplify. $10x^4 + 4x^2 + 3$

2. The vertical method:

   Add $(4x^7 + 6x^4 - 10 + 2x^5)$ and $(10x^5 - 5x^7 - 3x^4 + 12)$.
   Group like terms in columns in descending order.

$$\begin{array}{r} 4x^7 + 2x^5 + 6x^4 - 10 \\ + \ ^-5x^7 + 10x^5 - 3x^4 + 12 \\ \hline ^-x^7 + 12x^5 + 3x^4 + 2 \end{array}$$

   Simplify.

When subtracting polynomials, change the sign of each term in the second polynomial and then choose from one of the methods above to simplify.

Subtract. $(10x^7 - 4x^4 + 7x^2 + 2) - (12x^4 + 6x^7 + 8 - 5x^2)$

Change sign of each term in second polynomial. $10x^7 - 4x^4 + 7x^2 + 2 - 12x^4 - 6x^7 - 8 + 5x^2$

Choosing method 1 above, group like terms in descending order.

$(10x^7 - 6x^7) + (^-4x^4 - 12x^4) + (7x^2 + 5x^2) + (2 - 8)$

Simplify. $4x^7 - 16x^4 + 12x^2 - 6$

1. What is important to remember to do when subtracting polynomials?

   Change the sign of each term in the second polynomial.

Simplify.

2. $(7x - 5) - (3x + 7)$
   $4x - 12$

3. $(2x^2 - 3x + 1) - (5x^2 - 3x + 10)$
   $^-3x^2 - 9$

4. $(6x + 7) + (8x - 3)$
   $14x + 4$

5. $(9x^2 + 3x - 5) - (3x^2 + 4x - 10)$
   $6x^2 - x + 5$

6. $(8x^4 - 9x^3 + 2x) - (6x^4 + 7x^3 - 3x)$
   $2x^4 - 16x^3 + 5x$

7. $(x^3 - x^2 + 2) - (x^3 + x^2 + 5)$
   $^-2x^2 - 3$

© Carson-Dellosa • CD-104632

Name _____ (7.EE.A.1, HSA-APR.A.1)

Adding and Subtracting Polynomials

$(x^2 + 4x + 2) - (2x^2 + 7x - 6) = ^-x^2 - 3x + 8$

Add or subtract the polynomials by combining like terms.

1. $(4y^2 - 9y) - (^-5y^2 + 8y - 8) =$ **$9y^2 - 17y + 8$**

2. $(6x^2 + 2x + 6) - (4x^2 - 2x + 3) + (^-5x^2 + 5x + 6) =$ **$^-3x^2 + 9x + 9$**

3. $(^-2x^3 + 3x^2 + 9) + (^-8x^3 - 2x^2 + ^-4x) =$ **$^-10x^3 + x^2 - 4x + 9$**

4. $(2x^2 - 9x - 8) - (2x^3 - 7x^2 + ^-2) =$ **$^-2x^3 + 9x^2 - 9x - 6$**

5. $(4x^3 - 2x^2 - 12) + (6x^2 + 3x + 8) =$ **$4x^3 + 4x^2 + 3x - 4$**

6. $(3x^4 - 3x + 1) - (4x^3 - 4x - 8) =$ **$3x^4 - 4x^3 + x + 9$**

7. $(7x^2 - x - 5) - (3x^2 - 3x + 5) =$ **$4x^2 + 2x - 10$**

8. $(x^3 - x^2 + 3) - (3x^3 - x^2 + 7) =$ **$^-2x^3 - 4$**

9. $(^-2x^2 + 4x - 12) + (5x^2 - 5x) =$ **$3x^2 - x - 12$**

10. $(4x^3 - 5x^2 - 9) - (6x^3 - 5x - 4) - (5x^3 - 4x^2 - 10) =$ **$^-7x^3 - x^2 + 5x + 5$**

© Carson-Dellosa • CD-104632

Name _____ (7.EE.A.1, HSA-APR.A.1)

Adding and Subtracting Polynomials

Add or subtract the polynomials by combining like terms.

1. $(x^3 - 7x^4 + 2x^2 - 3x) + (4x^4 - 8x^3 + 5x) =$ **$-3x^4 - 7x^3 + 2x^2 + 2x$**

2. $(6x^2 + 9x + 4) - (^-7x^2 + 2x - 1) =$ **$13x^2 + 7x + 5$**

3. $(^-8x^2 + 2) + (7x - 5) =$ **$-8x^2 + 7x - 3$**

4. $(8x + 6) - (10x + 16) + (4x - 6) =$ **$2x - 16$**

5. $(6x^2 + 7x - 9) + (3x + 8) =$ **$6x^2 + 10x - 1$**

6. $(4x^2 + 2) + (3x^2 - 4x + 6) - (5x^2 + 10) =$ **$2x^2 - 4x - 2$**

7. $(2x^2 + 3x + 2) - (6x^3 - 3x^2 + 8) - (^-2x^3 + 9x^2 + 7) =$ **$-4x^3 - 4x^2 + 3x - 13$**

8. $(^-3x^2 - 4x^3 - 1) - (2x^3 - 7x - 9) - (2x^3 - 2x^2 - 3) =$ **$-8x^3 - x^2 + 7x + 11$**

9. $(^-6x^2 - 3x^3 + 4) + (^-7x^3 + 2x + 4) - (^-3x^3 + 5x^2 + 2) =$ **$-7x^3 - 11x^2 + 2x + 6$**

10. $(4x^2 + 6x + 3) + (3x^2 - 3x - 2) + (^-4x^2 + 3x - 9) =$ **$3x^2 + 6x - 8$**

11. $(9x^2 - 7x + ^-4) + (3x^3 - 4x + ^-5) + (^-4x^2 - 2x - 5) =$ **$3x^3 + 5x^2 - 13x - 14$**

12. $(3x^3 - 4x + 3) + (^-3x^3 + 5x - 8) =$ **$x - 5$**

13. $(^-5x^5 + 7x^3 - 4x + 2) + (7x^5 - 8x^4 + 9x^3) =$ **$2x^5 - 8x^4 + 16x^3 - 4x + 2$**

14. $(4x^2 - 8x + 2) - (^-5x^2 + 5x + 6) =$ **$9x^2 - 13x - 4$**

15. $(10x^3 - 8x^2 + 6x - 1) - (^-7x^3 + 2x - 4) =$ **$17x^3 - 8x^2 + 4x + 3$**

© Carson-Dellosa • CD-104632

Name _____ (6.EE.A.3, 7.EE.A.1, HSA-APR.A.1)

Multiplying Polynomials and Monomials

To multiply a multi-termed polynomial by a monomial, simply multiply each term in the polynomial by the monomial using the distributive property. Remember the properties of exponents when simplifying such problems.

Multiply.	$3x(3x^3 - 2x^2 + 8x - 7)$	
	$3x(3x^3) - 3x(2x^2) + 3x(8x) - 3x(7)$	Distribute $3x$ to each term.
	$9x^4 - 6x^3 + 24x^2 - 21x$	Multiply to simplify.

You may have to simplify an expression using the distributive property to multiply the polynomial by a monomial.

Simplify.	$4x^7 + 9x^4 - 7 + 2x^2(8x^5 - 5x^2)$	
	$4x^7 + 9x^4 - 7 + 2x^2(8x^5) - 2x^2(5x^2)$	Distribute the $2x$.
	$4x^7 + 9x^4 - 7 + 16x^7 - 10x^4$	Multiply to begin simplifying.
	$20x^7 - x^4 - 7$	Simplify.

Multiply each expression.

1. $6x(4x - 3)$
   $24x^2 - 18x$

2. $4x(x^2 - 6x + 3)$
   $4x^3 - 24x^2 + 12x$

3. $(^-3x^2 + 4x - 2)(^-4x^3)$
   $12x^5 - 16x^4 + 8x^3$

4. $x(^-5x + 2)$
   $^-5x^2 + 2x$

5. $^-3x(5x^2 - 2x - 6)$
   $^-15x^3 + 6x^2 + 18x$

6. $(^-x^2 + 2x - 1)(5x^3)$
   $^-5x^5 + 10x^4 - 5x^3$

7. $^-2x(^-4x - 8)$
   $8x^2 + 16x$

8. $^-x^2(7x^2 - x + 4)$
   $^-7x^4 + x^3 - 4x^2$

9. $^-5x^4(2x^2 - 8x + 6)$
   $^-10x^6 + 40x^5 - 30x^4$

10. $5y(y^2 - 3y + 1)$
    $5y^3 - 15y^2 + 5y$

11. $(3x^2 - 4x)(^-x)$
    $^-3x^3 + 4x^2$

12. $7x(x^3 - x^2)$
    $7x^4 - 7x^3$

© Carson-Dellosa • CD-104632

© Carson-Dellosa • CD-104632

Answer Key

Name _____ (6.EE.A.3, 7.EE.A.1, HSA-APR.A.1)

Multiplying Polynomials and Monomials

$4y(y - 3) = 4y^2 - 12y$

Use the distributive property to multiply the polynomials.

1. $a(a + 8) =$ **$a^2 + 8a$**
2. $5b(4b^3 - 6b^2 - 6) =$ **$20b^4 - 30b^3 - 30b$**
3. $3x(x - 3) =$ **$3x^2 - 9x$**
4. $4a(2a + 6) =$ **$8a^2 + 24a$**
5. $y(y - 7) =$ **$y^2 - 7y$**
6. $^-2x^2(5 - 3x + 3x^2 + 4x^3) =$ **$-10x^2 + 6x^3 - 6x^4 - 8x^5$**
7. $4b(3 - b) =$ **$12b - 4b^2$**
8. $2xy(2x - 3y) =$ **$4x^2y - 6xy^2$**
9. $^-5y^2(7y - 8y^2) =$ **$-35y^3 + 40y^4$**
10. $4x^2(3x^2 - x) =$ **$12x^4 - 4x^3$**
11. $x(x^2 + x + x) =$ **$x^3 + 2x^2$**
12. $3b(4b^3 - 12b^2 - 7) =$ **$12b^4 - 36b^3 - 21b$**
13. $(^-7x^3)(3x^2 - 1) =$ **$-21x^5 + 7x^3$**
14. $^-5ab(6a - 4b) =$ **$-30a^2b + 20ab^2$**
15. $3x(x - 3) =$ **$3x^2 - 9x$**
16. $^-3x^2(4x^2 - 3x + 3) =$ **$-12x^4 + 9x^3 - 9x^2$**
17. $^-4x^2(3x^3 + 8x^2 + {}^-9x) =$ **$-12x^5 - 32x^4 + 36x^3$**
18. $(3x^4 - 5x^2 - 4)(^-3x^3) =$ **$-9x^7 + 15x^5 + 12x^3$**

Name _____ (6.EE.A.3, 7.EE.A.1, HSA-APR.A.1)

Multiplying Polynomials and Monomials

Simplify each expression.

1. $^-4x^2 - 5x + 7 + 3(x^2 + 8x - 2)$ **$-x^2 + 19x + 1$**
2. $5x^2 - 3x(x - 7)$ **$2x^2 + 21x$**
3. $6x^2 + 4x + (7x - 3)2x$ **$20x^2 - 2x$**
4. $^-x^2 + 8x - 6 - 7(3x^2 - 5x + 9)$ **$-22x^2 + 43x - 69$**
5. $6x^2 + 3(x - 5) - 8x$ **$6x^2 - 5x - 15$**
6. $2x^2 + 7x + 6 - 5x(^-2x - 1)$ **$12x^2 + 12x + 6$**
7. $^-3x^2 - 3x(4x^2 - 5x + 7) - 8x^2$ **$-12x^3 + 4x^2 - 21x$**
8. $x^3 + (7x^2 - 9x - 1)x + 10x^2$ **$8x^3 + x^2 - x$**

Write and simplify an expression to find the area of each polygon. Remember that the area of a triangle can be found using $A = \frac{1}{2}bh$ and the area of a rectangle can be found using $A = lw$.

9. 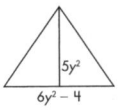

A = **$15y^4 - 10y^2$**

10. 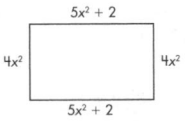

A = **$20x^4 + 8x^2$**

11. 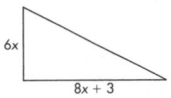

A = **$24x^2 + 9x$**

12. 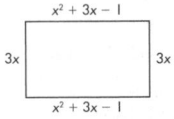

A = **$3x^3 + 9x^2 - 3x$**

Name _____ (7.EE.A.1, HSA-SSE.A.2, HSA-APR.A.1, HSA-APR.C.4)

Multiplying Binomials

The easiest way to multiply binomials is to use what is called the FOIL method. This method multiplies the first terms of the binomials (F), multiplies the outer terms of the binomials (O), multiplies the inner terms of the binomials (I), multiplies the last terms of the binomials (L), and simplifies by adding like terms.

1. Multiply $(3x - 2)(4x + 3)$

$3x(4x) = 12x^2$	Multiply first terms.	F
$3x(3) = 9x$	Multiply outer terms.	O
$^-2(4x) = {}^-8x$	Multiply inner terms.	I
$^-2(3) = {}^-6$	Multiply last terms.	L
$12x^2 + 9x - 8x - 6$	Simplify by combining like terms.	
$12x^2 + x - 6$	Final product	

2. Multiply $(5x + 2)(7x - 3)$

$5x(7x) + 5x(^-3) + 2(7x) + 2(^-3)$
F O I L

$35x^2 - 15x + 14x - 6$	Multiply.
$35x^2 - x - 6$	Final product

Multiply.

1. $(x + 7)(x - 5)$ **$x^2 + 2x - 35$**
2. $(x - 5)(3x + 7)$ **$3x^2 - 8x - 35$**
3. $(9x - 1)(6x + 2)$ **$54x^2 + 12x - 2$**
4. $(7x - 8)(8x - 7)$ **$56x^2 - 113x + 56$**
5. $(x + 2)(x + 3)$ **$x^2 + 5x + 6$**
6. $(2b - 8)(3b - 7)$ **$6b^2 - 38b + 56$**
7. $(3x + 5)(^-4x - 7)$ **$-12x^2 - 41x - 35$**
8. $(^-3x - 9)(^-x - 6)$ **$3x^2 + 27x + 54$**
9. $(x - 10)(x + 1)$ **$x^2 - 9x - 10$**
10. $(2x - 3)(5x + 4)$ **$10x^2 - 7x - 12$**
11. $(^-5x + 6)(x - 2)$ **$-5x^2 + 16x - 12$**
12. $(^-4x + 4)(5x + 8)$ **$-20x^2 - 12x + 32$**
13. $(5b - 2)(3b + 2)$ **$15b^2 + 4b - 4$**
14. $(3x + y)(3x - 2y)$ **$9x^2 - 3xy - 2y^2$**
15. $(4a + 1)(3a - 3)$ **$12a^2 - 9a - 3$**
16. $(x - y)(2x + 2y)$ **$2x^2 - 2y^2$**

Name _____ (7.EE.A.1, HSA-SSE.A.2, HSA-APR.A.1, HSA-APR.C.4)

Multiplying Binomials

Rules: $(a + b)(a - b) = a^2 - b^2$
$(a + b)^2 = a^2 + 2ab + b^2$
$(a - b)^2 = a^2 - 2ab + b^2$

Multiply.

1. $(2x + y)(2x - y) =$ **$4x^2 - y^2$**
2. $(b - 5)(b + 5) =$ **$b^2 - 25$**
3. $(x - y)(2x + 2y) =$ **$2x^2 - 2y^2$**
4. $(x - 4y)^2 =$ **$x^2 - 8xy + 16y^2$**
5. $(7x - 3y)(7x + 3y) =$ **$49x^2 - 9y^2$**
6. $(x + 6)(x - 6) =$ **$x^2 - 36$**
7. $(7x + y)(7x - y) =$ **$49x^2 - y^2$**
8. $(2x - 6y)^2 =$ **$4x^2 - 24xy + 36y^2$**
9. $(3a - b)(3a + b) =$ **$9a^2 - b^2$**
10. $(x + 2)(x - 2) =$ **$x^2 - 4$**
11. $(12b - 5)(12b + 5) =$ **$144b^2 - 25$**
12. $(2x - 3y)^2 =$ **$4x^2 - 12xy + 9y^2$**
13. $(7x - 5y)^2 =$ **$49x^2 - 70xy + 25y^2$**
14. $(3x + 13)(3x - 2) =$ **$9x^2 + 33x - 26$**
15. $(c + 2d)(c - 2d) =$ **$c^2 - 4d^2$**
16. $(3b + 6)(3b - 6) =$ **$9b^2 - 36$**
17. $(2b + 5a)^2 =$ **$4b^2 + 20ab + 25a^2$**
18. $(2x + 3v)^2 =$ **$4x^2 + 12xv + 9v^2$**
19. $(12 + b)(12 - b) =$ **$144 - b^2$**
20. $(5x + 7)^2 =$ **$25x^2 + 70x + 49$**

Answer Key

Name _____ 7.EE.A.1, HSA-SSE.A.2, HSA-APR.A.1, HSA-APR.C.4

Multiplying Binomials

Multiply.

1. $(4b^2 - 4)(4b^2 + 4) =$ **$16b^4 - 16$**
2. $(3b - 3c)^2 =$ **$9b^2 - 18bc + 9c^2$**
3. $(x - 2y)^2 =$ **$x^2 - 4xy + 4y^2$**
4. $(-5x^2 + 3)(-5x^2 - 3) =$ **$25x^4 - 9$**
5. $(7b^2 - 3c)^2 =$ **$49b^4 - 42b^2c + 9c^2$**
6. $(4c + 9d)^2 =$ **$16c^2 + 72cd + 81d^2$**
7. $(5x^2 - 5y)^2 =$ **$25x^4 - 50x^2y + 25y^2$**
8. $(x - yz)(x + yz) =$ **$x^2 - y^2z^2$**
9. $(-4x + 3y)^2 =$ **$16x^2 - 24xy + 9y^2$**
10. $(4m^2 - 2n)^2 =$ **$16m^4 - 16m^2n + 4n^2$**
11. $(x^2 - 7y)^2 =$ **$x^2 - 14x^2y + 49y^2$**
12. $(4x^2 - 4y^2)^2 =$ **$16x^4 - 32x^2y^2 + 16y^4$**
13. $(8x^2 - 12)(8x^2 + 12) =$ **$64x^4 - 144$**
14. $(2b^2 - 2c^2)^2 =$ **$4b^4 - 8b^2c^2 + 4c^4$**
15. $(4a + b)^2 =$ **$16a^2 + 8ab + b^2$**
16. $(x^2 - 8x)(x^2 + 8x) =$ **$x^4 - 64x^2$**
17. $(2x^2 - y^2)(2x^2 + y^2) =$ **$4x^4 - y^4$**
18. $(3x^2 - x)(3x^2 + x) =$ **$9x^4 - x^2$**
19. $(3a - 7b)^2 =$ **$9a^2 - 42ab + 49b^2$**
20. $(-6x + 3y)^2 =$ **$36x^2 - 36xy + 9y^2$**
21. $(4x^2 - 4y^2)(4x^2 + 4y^2) =$ **$16x^4 - 16y^4$**
22. $(-2x^3 + 4)(2x^2 + 5) =$ **$-4x^5 - 10x^3 + 8x^2 + 20$**

© Carson-Dellosa • CD-104632

Name _____ 7.EE.A.1, HSA-SSE.A.2

Factoring Polynomials

To factor a polynomial, write the polynomial as a product of other polynomials.
For example, $3x^2 - 6x$ can be written as $3x(x - 2)$.
$3x$ is the Greatest Common Factor (GCF) of $3x^2$ and $6x$.
$3x$ is a Common Monomial Factor of the terms of the binomial.
$x - 2$ is a Binomial Factor of $3x^2 - 6$.

To factor a trinomial in the form $x^2 + bx + c$, write it as the product of two binomials.
For example, $a^2 - 9a + 14$ can be written as $(a - 7)(a - 2)$.

Look for two numbers that are the product of c and whose sum is b.
If c and b are both positive, the factors will both be positive.
If c is positive and b is negative, the factors will both be negative.
If c is negative, the factors will have opposite signs.

Factor.

1. $3x^2 - 12x^3 =$ **$3x^2(1 - 4x)$**
2. $2x^3 - x^4 =$ **$x^3(2 - x)$**
3. $3a^5 - a^3 =$ **$a^3(3a^2 - 1)$**
4. $x^5 + 2x^2 =$ **$x^2(x^3 + 2)$**
5. $24b^2 + 16b =$ **$8b(3b + 2)$**
6. $5x^3 - 7x^2 =$ **$x^2(5x - 7)$**
7. $2x^3 + 6x^2 =$ **$2x^2(x + 3)$**
8. $x^3 - 5x^2 =$ **$x^2(x - 5)$**
9. $x^2 + x - 90 =$ **$(x + 10)(x - 9)$**
10. $y^2 - 13y + 42 =$ **$(y - 6)(y - 7)$**
11. $x^2 - x - 6 =$ **$(x - 3)(x + 2)$**
12. $x^2 - 9x + 18 =$ **$(x - 3)(x - 6)$**
13. $b^2 - 4b - 45 =$ **$(b - 9)(b + 5)$**
14. $y^2 - 12y + 36 =$ **$(y - 6)(y - 6)$**

© Carson-Dellosa • CD-104632

Name _____ 7.EE.A.1, HSA-SSE.A.2

Factoring Polynomials

$a^2 - 9a + 14 = (a - 7)(a - 2)$
$2x^2 - 5x - 12 = (2x + 3)(x - 4)$

Factor.

1. $p^2 + 13p + 42 =$ **$(p + 7)(p + 6)$**
2. $c^2 + c - 30 =$ **$(c + 6)(c - 5)$**
3. $x^2 + 15xy + 44y^2 =$ **$(x + 4y)(x + 11y)$**
4. $x^2 - 13x + 12 =$ **$(x - 12)(x - 1)$**
5. $x^2 - 13x + 30 =$ **$(x - 10)(x - 3)$**
6. $x^2 - 8x + 16 =$ **$(x - 4)(x - 4)$**
7. $x^2 - 12xy + 32y^2 =$ **$(x - 8y)(x - 4y)$**
8. $x^2 + 14x + 49 =$ **$(x + 7)(x + 7)$**
9. $n^2 + 6n - 16 =$ **$(n + 8)(n - 2)$**
10. $x^2 + 12x + 35 =$ **$(x + 7)(x + 5)$**
11. $c^2 - 10c + 21 =$ **$(c - 7)(c - 3)$**
12. $x^2 + 6x - 40 =$ **$(x + 10)(x - 4)$**
13. $2x^2 + 9x + 10 =$ **$(2x + 5)(x + 2)$**
14. $4x^2 - 18x + 20 =$ **$2(2x - 5)(x - 2)$**
15. $3x^2 - 10x + 7 =$ **$(3x - 7)(x - 1)$**
16. $3x^2 - 5x - 12 =$ **$(3x + 4)(x - 3)$**
17. $3x^2 - 4x - 32 =$ **$(3x + 8)(x - 4)$**
18. $5x^2 + 25x + 30 =$ **$5(x + 3)(x + 2)$**

© Carson-Dellosa • CD-104632

Name _____ 7.EE.A.1, HSA-SSE.A.2

Factoring Polynomials

Factor.

1. $x^2 + 4x + 4 =$ **$(x + 2)(x + 2)$**
2. $2x^2 + 13x + 6 =$ **$(2x + 1)(x + 6)$**
3. $x^2 + 7x + 12 =$ **$(x + 3)(x + 4)$**
4. $6x^2 - 21x - 12 =$ **$3(2x + 1)(x - 4)$**
5. $x^2 - 8x + 15 =$ **$(x - 3)(x - 5)$**
6. $9x^2 - 9x - 28 =$ **$(3x + 4)(3x - 7)$**
7. $x^2 - 15x + 56 =$ **$(x - 7)(x - 8)$**
8. $15x^2 + 11x - 14 =$ **$(5x + 7)(3x - 2)$**
9. $x^2 + 18x + 45 =$ **$(x + 15)(x + 3)$**
10. $10x^2 - 28x - 6 =$ **$2(x - 3)(5x + 1)$**
11. $x^2 + 4x - 32 =$ **$(x + 8)(x - 4)$**
12. $7x^2 + 17x + 6 =$ **$(7x + 3)(x + 2)$**
13. $x^2 + 2xy - 63y^2 =$ **$(x + 9y)(x - 7y)$**
14. $2x^2 - 2x - 40 =$ **$2(x - 5)(x + 4)$**
15. $x^2 - 14x - 72 =$ **$(x + 4)(x - 18)$**
16. $11x^2 - 122x + 11 =$ **$(11x - 1)(x - 11)$**
17. $x^2 - xy - 2y^2 =$ **$(x - 2y)(x + y)$**
18. $2x^2 + 7x + 3 =$ **$(2x + 1)(x + 3)$**
19. $x^2 + 16x + 28 =$ **$(x + 14)(x + 2)$**
20. $12x^2 + 9x - 3 =$ **$3(4x - 1)(x + 1)$**

© Carson-Dellosa • CD-104632

© Carson-Dellosa • CD-104632

Answer Key

Name _____ (7.EE.A.1, HSA-SSE.A.2)

Factoring Polynomials—Special Cases

Factoring Trinomials That Are Quadratic in Form
$x^4 - x^2 - 12 = (x^2)^2 - (x^2) - 12 = (x^2 - 4)(x^2 + 3)$

Factoring the Difference of Two Squares
Rule: $a^2 - b^2 = (a + b)(a - b)$ Example: $x^2 - 49 = (x + 7)(x - 7)$

Factoring Perfect Square Trinomials
Rules: $a^2 + 2ab + b^2 = (a + b)^2$ $a^2 - 2ab + b^2 = (a - b)^2$
Examples: $9x^2 + 6x + 1 = (3x + 1)^2$ $x^2 - 6x + 9 = (x - 3)^2$

Factoring the Sum or Difference of Two Cubes
Rules: $x^3 + y^3 = (x + y)(x^2 - xy + y^2)$ $x^3 - y^3 = (x - y)(x^2 + xy + y^2)$
Examples: $x^3 + 8 = (x + 2)(x^2 - 2x + 4)$ $x^3 - 8 = (x - 2)(x^2 + 2x + 4)$

Factor.

1. $2x^2 - 5x - 12 =$ **$(2x + 3)(x - 4)$**
2. $2x^4 + 16x^2 + 30 =$ **$(2x^2 + 10)(x^2 + 3)$**
3. $x^4 - 8x^2 + 15 =$ **$(x^2 - 3)(x^2 - 5)$**
4. $7x^4 - 11x^2 - 6 =$ **$(7x^2 + 3)(x^2 - 2)$**
5. $a^2 - 4 =$ **$(a - 2)(a + 2)$**
6. $b^2 - 9 =$ **$(b - 3)(b + 3)$**
7. $1 - 9x^2 =$ **$(1 - 3x)(1 + 3x)$**
8. $x^2 - 25 =$ **$(x - 5)(x + 5)$**
9. $x^2 + 14x + 49 =$ **$(x + 7)^2$**
10. $x^2 - 2x + 1 =$ **$(x - 1)^2$**
11. $c^2 - 6c + 9 =$ **$(c + 3)^2$**
12. $x^2 - 4xy + 4y^2 =$ **$(x - 2y)^2$**
13. $64x^3 + 1 =$ **$(4x + 1)(16x^2 - 4x + 1)$**
14. $8x^3 + 27 =$ **$(2x + 3)(4x^2 - 6x + 9)$**

Name _____ (7.EE.A.1, HSA-SSE.A.2)

Factoring Polynomials—Special Cases

Factor.

1. $2x^4 - 7x^2 - 15 =$ **$(2x^2 + 3)(x^2 - 5)$**
2. $y^4 + 6y^2 - 16 =$ **$(y^2 - 2)(y^2 + 8)$**
3. $8x^4 - 23x^2 - 3 =$ **$(8x^2 + 1)(x^2 - 3)$**
4. $6a^6 - 5a^3b^3 - 25b^6 =$ **$(3a^3 + 5b^3)(2a^3 - 5b^3)$**
5. $3x^4 + 20x^2 + 33 =$ **$(3x^2 + 11)(x^2 + 3)$**
6. $4x^4y^4 - 2x^2y^2 - 56 =$ **$(2x^2y^2 + 7)(2x^2y^2 - 8)$**
7. $y^2 - 81 =$ **$(y - 9)(y + 9)$**
8. $c^2 - 16 =$ **$(c - 4)(c + 4)$**
9. $a^2 - 49 =$ **$(a - 7)(a + 7)$**
10. $49x^2 - 16y^4 =$ **$(7x - 4y^2)(7x + 4y^2)$**
11. $16x^2 - 121 =$ **$(4x - 11)(4x + 11)$**
12. $25 - x^2y^2 =$ **$(5 - xy)(5 + xy)$**
13. $a^2 + 12ab + 36b^2 =$ **$(a + 6b)^2$**
14. $49x^2 + 28x + 4 =$ **$(7x + 2)^2$**
15. $x^2 + 14x + 49 =$ **$(x + 7)^2$**
16. $c^2 - 20c + 100 =$ **$(c - 10)^2$**
17. $x^3 - 1000 =$ **$(x - 10)(x^2 + 10x + 100)$**
18. $1 - 64y^3 =$ **$(1 - 4y)(1 + 4y + 16y^2)$**
19. $x^3 + 125 =$ **$(x + 5)(x^2 - 5x + 25)$**
20. $x^3 - 27 =$ **$(x - 3)(x^2 + 3x + 9)$**

Name _____ (7.EE.A.1, HSA-SSE.A.2)

Factoring Polynomials—Special Cases

Factor.

1. $y^4 - y^2 - 12 =$ **$(y^2 + 3)(y^2 - 4)$**
2. $x^4y^4 - 19x^2y^2 + 34 =$ **$(x^2y^2 - 2)(x^2y^2 - 17)$**
3. $2x^4y^4 - 17x^2y^2 - 30 =$ **$(2x^2y^2 + 3)(x^2y^2 - 10)$**
4. $x^4y^4 - 8x^2y^2 + 12 =$ **$(x^2y^2 - 6)(x^2y^2 - 2)$**
5. $64 - x^4y^4 =$ **$(8 - x^2y^2)(8 + x^2y^2)$**
6. $y^2 - 64 =$ **$(y - 8)(y + 8)$**
7. $81x^2 - 4 =$ **$(9x - 2)(9x + 2)$**
8. $16 - 81x^2 =$ **$(4 - 9x)(4 + 9x)$**
9. $x^2y^2 - 121 =$ **$(xy - 11)(xy + 11)$**
10. $49x^2 - 36 =$ **$(7x - 6)(7x + 6)$**
11. $y^2 - 22y + 121 =$ **$(y - 11)^2$**
12. $25a^2 - 40ab + 16b^2 =$ **$(5a - 4b)^2$**
13. $9x^2 - 12x + 4 =$ **$(3x - 2)^2$**
14. $16x^2 - 40x + 25 =$ **$(4x - 5)^2$**
15. $9x^2 + 12x + 4 =$ **$(3x + 2)^2$**
16. $x^2 - 14x + 49 =$ **$(x - 7)^2$**
17. $27x^3 - 64 =$ **$(3x - 4)(9x^2 + 12x + 16)$**
18. $27x^3 - 27 =$ **$27(x - 1)(x^2 + x + 1)$**
19. $125a^3 - 8b^3 =$ **$(5a - 2b)(25a^2 + 10ab + 4b^2)$**
20. $64x^3 - y^3 =$ **$(4x - y)(16x^2 + 4xy + y^2)$**
21. $125x^3 - 64y^3 =$ **$(5x - 4y)(25x^2 + 20xy + 16y^2)$**
22. $64x^3 + 27 =$ **$(4x + 3)(16x^2 - 12x + 9)$**

Name _____ (HSA-SSE.B.3a, HSA-REI.B.4b)

Solving Equations by Factoring

The **Multiplication Property of Zero**: The product of a number and zero is zero.

The **Principle of Zero Products**: If the product of two factors is zero, then at least one of the factors must be zero. This principle is used in solving equations.

Solve: $(x - 4)(x - 5) = 0$
$x - 4 = 0$ $x - 5 = 0$
$x = 4$ $x = 5$
The solutions are 4 and 5.

If $(x - 4)(x - 5) = 0$, then $(x - 4) = 0$ or $(x - 5) = 0$.

$x = 4$	$x = 5$
$(4 - 4)(4 - 5) = 0$	$(5 - 4)(5 - 5) = 0$
$(0)(-1) = 0$	$(1)(0) = 0$
$0 = 0$	$0 = 0$

Write the solutions for each variable.

1. $x(x + 6) = 0$ **0, −6**
2. $b^2 - 81 = 0$ **−9, 9**
3. $(27 - y)(y - 2) = 0$ **2, 27**
4. $z^2 - 1 = 0$ **−1, 1**
5. $(y - 4)(y - 8) = 0$ **4, 8**
6. $y(y - 11) = 0$ **0, 11**
7. $8b^2 - 32 = 0$ **2, −2**
8. $x^2 - x - 6 = 0$ **−2, 3**
9. $x^2 - 4x - 21 = 0$ **−3, 7**
10. $m^2 - 144 = 0$ **−12, 12**
11. $(y + 5)(y + 6) = 0$ **−5, −6**
12. $(2x + 4)(x + 7) = 0$ **−7, −2**
13. $z^2 - 9 = 0$ **−3, 3**
14. $10x^2 - 10x = 0$ **1, 0**
15. $2x^2 - 6x = x - 3$ **$\frac{1}{2}$, 3**
16. $4y(3y - 2) = 0$ **0, $\frac{2}{3}$**
17. $(4y - 1)(y + 2) = 0$ **$\frac{1}{4}$, −2**
18. $x^2 - 5x + 6 = 0$ **2, 3**

Answer Key

Name _____ HSA-SSE.B.3a, HSA-REI.B.4b

Solving Equations by Factoring

$x^2 - 8x = {}^-16$	$x - 4 = 0$
$x^2 - 8x + 16 = 0$	$x = 4$
$(x - 4)(x - 4) = 0$	The solution is 4.

Solve by factoring.

1. $y^2 + 9y = 0$ **$-9, 0$**
2. $x - 16 = x(x - 7)$ **4**
3. $x - 6 = x(x - 4)$ **2, 3**
4. $x^2 + 7x = 0$ **$0, -7$**
5. $x^2 - 4x = 0$ **0, 4**
6. $x + 8 = x(x + 3)$ **$-4, 2$**
7. $y^2 - y - 6 = 0$ **$-2, 3$**
8. $a^2 - 36 = 0$ **$-6, 6$**
9. $y^2 + 15 = 8y$ **3, 5**
10. $a^2 - 7a = {}^-12$ **3, 4**
11. $y^2 + 36y = 0$ **$0, -36$**
12. $3u^2 - 12u - 15 = 0$ **$-1, 5$**
13. $y^2 - 8y + 12 = 0$ **2, 6**
14. $5a^2 + 25a = 0$ **$-5, 0$**
15. $6x^2 + 18x = 0$ **$-3, 0$**
16. $2x^2 + x = 6$ **$-2, \frac{3}{2}$**
17. $x^2 - 5x - 6 = 0$ **$-1, 6$**
18. $4x^2 + 16x = 0$ **$0, -4$**
19. $3x^2 - 9x = 0$ **0, 3**
20. $y^2 + 5y - 6 = 0$ **$-6, 1$**

© Carson-Dellosa • CD-104632

Name _____ HSA-SSE.B.3a, HSA-REI.B.4b

Solving Equations by Factoring

Solve by factoring.

1. $a^2 - 8a = 0$ **0, 8**
2. $x^2 = 3x + 4$ **$-1, 4$**
3. $4a^2 + 15a - 4 = 0$ **$\frac{1}{4}, -4$**
4. $x^2 - x - 6 = 0$ **$-2, 3$**
5. $3x^2 - 13x + 4 = 0$ **$\frac{1}{3}, 4$**
6. $6x^2 = 23x + 18$ **$-\frac{2}{3}, \frac{9}{2}$**
7. $x^2 + 7x + 12 = 0$ **$-4, -3$**
8. $x^2 + 5x - 6 = 0$ **$-6, 1$**
9. $x^2 = 6x + 7$ **$-1, 7$**
10. $x^2 = 10x - 25$ **5**
11. $x^2 + 3x - 10 = 0$ **$-5, 2$**
12. $x^2 - 6x + 9 = 0$ **3**
13. $y^2 - 3y + 2 = 0$ **1, 2**
14. $2x^2 - 9x + 9 = 0$ **$\frac{3}{2}, 3$**
15. $r^2 - 15r = 16$ **$-1, 16$**
16. $x^2 + 7x + 10 = 0$ **$-5, -2$**
17. $3x^2 - 2x - 8 = 0$ **$-\frac{4}{3}, 2$**
18. $2a^2 + 4a - 6 = 0$ **$-3, 1$**
19. $x^2 + 3x - 4 = 0$ **$-4, 1$**
20. $4a^2 + 9a + 2 = 0$ **$-2, -\frac{1}{4}$**
21. $9x^2 = 18x + 0$ **0, 2**
22. $2x^2 = 9x + 5$ **$-\frac{1}{2}, 5$**

© Carson-Dellosa • CD-104632

Name _____ 7.EE.B.4, HSA-SSE.B.3a, HSA-REI.B.4b

Writing and Solving Quadratic Equations

Problem solving often involves quadratic equations.

The square of a number is 8 less than 6 times the number. Find the numbers that make the sentence true.

$x^2 = 6x - 8$	First, set up the equation.
$x^2 - 6x + 8 = 0$	Put the equation in standard form.
$(x - 2)(x - 4) = 0$	Factor.
$x - 2 = 0$ or $x - 4 = 0$	Solve for x.
$x = 2$ or $x = 4$	So, the numbers are 2 and 4.

The product of a number and 10 more than 4 times the number is 50. Find the numbers that make the sentence true.

$x(4x + 10) = 50$	First, set up the equation.
$4x^2 + 10x - 50 = 0$	Put the equation in standard form.
$2(2x - 5)(x + 5) = 0$	Factor.
Since $2 \neq 0$, then $2x - 5 = 0$ or $x + 5 = 0$	Solve for x.
$x = \frac{5}{2}$ and $x = {}^-5$	So, the numbers are $\frac{5}{2}$ and $^-5$.

Write an equation for each word problem and solve it.

1. Six less than 3 times a number is 12 less than twice the number.

   Equation **$3x - 6 = 2x - 12$** **Solution** **-6**

2. The square of a number is 8 more than twice the number.

   Equation **$x^2 = 2x + 8$** **Solution** **$4, -2$**

3. The sum of the square of a number and 4 times the number is 12.

   Equation **$x^2 + 4x = 12$** **Solution** **$-6, 2$**

4. Twice the square of a number is 4 more than twice the number.

   Equation **$2x^2 = 2x + 4$** **Solution** **$-1, 2$**

5. Three times the square of a number is 5 less than 16 times the number.

   Equation **$3x^2 = 16x - 5$** **Solution** **$\frac{1}{3}, 5$**

© Carson-Dellosa • CD-104632

Name _____ 7.EE.B.4, HSA-SSE.B.3a, HSA-REI.B.4b

Writing and Solving Quadratic Equations

The length of a rectangle is 3 inches longer than the width. The area of the rectangle is 40 square inches. Find the length and width of the rectangle.

Width of rectangle: w	
Length of rectangle: $w + 3$	
$A = lw$	
$40 = (w + 3)(w)$	Since the width cannot be a negative number, the width is 5.
$40 = w^2 + 3w$	
$0 = w^2 + 3w - 40$	
$0 = (w + 8)(w - 5)$	$l = 5 + 3 = 8$
$w + 8 = 0$ or $w - 5 = 0$	
$w = {}^-8$ $w = 5$	The length is 8 inches, and the width is 5 inches.

Write an equation for each word problem and solve it.

1. The area of a square is 121 m². Find the length of the sides of the square.

   Equation **$s^2 = 121$** **Solution** **$s = 11$ m**

2. The area of a rectangle is 72 m². Its length is twice its width. Find the length and width of the rectangle.

   Equation **$2w(w) = 72$** **Solution** **$l = 12$ m, $w = 6$ m**

3. The area of a rectangle is 36 cm². Its width is 4 times its length. Find the length and width of the rectangle.

   Equation **$(l)4l = 36$** **Solution** **$l = 3$ cm, $w = 12$ cm**

4. The width of a rectangle is 5 more than twice its length. The area of the rectangle is 33 in.². Find the dimensions of the rectangle.

   Equation **$(2l + 5)(l) = 33$** **Solution** **$l = 3$ in, $w = 11$ in.**

5. The length of a rectangle is 4 more than twice its width. The area of the rectangle is 96 ft.². Find its dimensions.

   Equation **$w(2w + 4) = 96$** **Solution** **$l = 16$ ft, $w = 6$ ft.**

© Carson-Dellosa • CD-104632

© Carson-Dellosa • CD-104632

Answer Key

Name _____ (7.EE.B.4, HSA-SSE.B.3a, HSA-REI.B.4b)

Writing and Solving Quadratic Equations

Write an equation for each word problem and solve it.

1. The length of a rectangle is 5 inches longer than the width. The area of the rectangle is 50 square inches. Find the length and width.

   Equation $w(w + 5) = 50$ **Solution** $l = 10$ in, $w = 5$ in.

2. The length of a rectangle is 2 more than 8 times its width. Its area is 36 cm². Find the width and length.

   Equation $w(8w + 2) = 36$ **Solution** $l = 18$ cm, $w = 2$ cm

3. The sum of a number and its square is 42. Find the numbers.

   Equation $x + x^2 = 42$ **Solution** $x = -7, 6$

4. The sum of a number and its square is 56. Find the numbers.

   Equation $x + x^2 = 56$ **Solution** $x = -8, 7$

5. The length of a rectangle is twice its width. Its area is 32 m². Find the length and width.

   Equation $w(2w) = 32$ **Solution** $l = 8$ m, $w = 4$ m

6. The square of a number is 80 more than 2 times the number. Find the numbers.

   Equation $x^2 = 2x + 80$ **Solution** $x = 10, -8$

7. The square of a number is 48 more than 2 times the number. Find the numbers.

   Equation $x^2 = 2x + 48$ **Solution** $x = 8, -6$

8. The width of a rectangle is 2/3 the length. The area is 54 in². Find the length and width.

   Equation $\frac{2}{3}l^2 = 54$ **Solution** $l = 9$ in, $w = 6$ in.

Name _____ (HSA-APR.D.6)

Dividing Polynomials

Simplify using long division: $(x^2 + 6x + 5) \div (x + 1)$

$$\frac{(x^2 + 6x + 5)}{(x + 1)} \qquad \begin{array}{r} x + 5 \\ x + 1\overline{)x^2 + 6x + 5} \\ -\ x^2 + 1x \\ \hline 5x + 5 \\ -\ 5x + 5 \\ \hline 0 \end{array}$$

Divide by using long division.

1. $(x^2 + 5x + 6) \div (x + 3) =$ **$x + 2$**
2. $(x^2 + 4x - 21) \div (x - 3) =$ **$x + 7$**
3. $(x^2 - 3x - 40) \div (x + 5) =$ **$x - 8$**
4. $(x^2 - x - 42) \div (x + 6) =$ **$x - 7$**
5. $(x^2 - 8x + 16) \div (x - 4) =$ **$x - 4$**
6. $(x^2 + 2x - 35) \div (x + 7) =$ **$x - 5$**
7. $(x^2 - 6x + 9) \div (x - 3) =$ **$x - 3$**
8. $(x^2 + 5x + 4) \div (x + 1) =$ **$x + 4$**
9. $(x^2 + 7x + 10) \div (x + 2) =$ **$x + 5$**
10. $(x^2 + 9x + 8) \div (x + 8) =$ **$x + 1$**

Name _____ (HSA-APR.D.6)

Dividing Polynomials

Simplify using synthetic division: $(2x^3 + 3x^2 - 4x + 8) \div (x + 3)$

$$\begin{array}{r|rrrr} -3 & 2 & 3 & -4 & 8 \\ & & -6 & 9 & -15 \\ \hline & 2 & -3 & 5 & -7 \end{array}$$

$$= 2x^2 - 3x + 5 - \frac{7}{x + 3}$$

Divide by using synthetic division.

1. $(x^2 + x - 2) \div (x + 2) =$ **$x - 1$**
2. $(5x^2 - 12x - 9) \div (x - 3) =$ **$5x + 3$**
3. $(3x^2 - 5) \div (x - 1) =$ **$3x + 3 - \frac{2}{(x - 1)}$**
4. $(3x^3 + 8x^2 + 9x + 10) \div (x + 2) =$ **$3x^2 + 2x + 5$**
5. $(x^3 - 4x^2 - 36x - 16) \div (x + 4) =$ **$x^2 - 8x - 4$**
6. $(3x^2 - 7x + 6) \div (x - 3) =$ **$3x + 2 + \frac{12}{(x - 3)}$**
7. $(4x^2 + 9x + 6) \div (x + 1) =$ **$4x + 5 + \frac{1}{(x + 1)}$**
8. $(4x^2 + 23x + 28) \div (x + 4) =$ **$4x + 7$**
9. $(3x^2 + 19x + 20) \div (x + 5) =$ **$3x + 4$**
10. $(x^2 + 14x + 45) \div (x + 5) =$ **$x + 9$**

Name _____ (HSA-APR.D.6)

Dividing Polynomials

Divide.

1. $\frac{x^2 + 9x + 8}{x - 1} =$ **$x - 8$**
2. $\frac{2x^2 - 5x - 3}{x - 3} =$ **$2x + 1$**
3. $\frac{4x^2 - 7x - 2}{4x + 1} =$ **$x - 2$**
4. $\frac{2x^2 + x - 3}{x - 1} =$ **$2x + 3$**
5. $\frac{21x^2 + 22x - 8}{3x + 4} =$ **$7x - 2$**
6. $\frac{3x^2 - 27}{x + 3} =$ **$3x - 9$**
7. $\frac{9x^2 - 27x - 36}{x - 4} =$ **$9x + 9$**
8. $\frac{5x^2 + 43x - 18}{x + 9} =$ **$5x - 2$**
9. $\frac{x^2 - 4x - 45}{x + 5} =$ **$x - 9$**
10. $\frac{2x^2 - x - 21}{x + 3} =$ **$2x - 7$**
11. $(3x^3 - 13x^2 - 13x + 15) \div (x - 5) =$ **$3x^2 + 2x - 3$**
12. $(2x^3 - 12x^2 + 5x - 27) \div (x - 6) =$ **$2x^2 + 5 + \frac{3}{(x - 6)}$**
13. $(3x^2 - 75) \div (x - 5) =$ **$3x + 15$**
14. $(2x^2 + 7x - 10) \div (x + 1) =$ **$2x + 5 - \frac{15}{(x + 1)}$**

Answer Key

Name _____ HSA-APR.D.7

Operations with Rational Expressions

Add or subtract:

$$\frac{7x-12}{2x^2+5x-12} - \frac{3x-6}{2x^2+5x-12} = \frac{(7x-12)-(3x-6)}{2x^2+5x-12} = \frac{4x-6}{(2x-3)(x+4)} = \frac{2\cancel{(2x-3)}}{\cancel{(2x-3)}(x+4)} = \frac{2}{x+4}$$

Multiply:

$$\frac{4x+16}{12x+48} \cdot \frac{(x-4)}{2x} = \frac{\cancel{4(x+4)}(x-4)}{\cancel{24x(x+4)}} = \frac{x-4}{6x}$$

Divide:

$$\frac{x+2}{x+3} \div \frac{x^2-4}{x-2} = \frac{\cancel{x+2}}{x+3} \cdot \frac{\cancel{x-2}}{\cancel{(x+2)}\cancel{(x-2)}} = \frac{1}{x+3}$$

Simplify.

1. $\frac{x^2+x-6}{x+1} \cdot \frac{x+1}{x^2-9} = \mathbf{\frac{(x-2)}{(x-3)}}$
2. $\frac{x^2-1}{x^2-2x-3} \cdot \frac{x+4}{6x-6} = \mathbf{\frac{(x+4)}{6(x-3)}}$
3. $\frac{x-7}{x+2} \div \frac{x^2-49}{x^2+9x+14} = \mathbf{1}$
4. $\frac{x+2}{4x(x-6)} \div \frac{x^2-4}{8x(x-6)} = \mathbf{\frac{2}{(x-2)}}$
5. $\frac{2}{x+2} + \frac{6x}{x^2-4} = \mathbf{\frac{8x-4}{[(x-2)(x+2)]}}$
6. $\frac{2}{4xy} + \frac{14}{3xy} - \frac{9}{2xy} = \mathbf{\frac{11}{12xy}}$

Name _____ HSA-APR.D.7

Operations with Rational Expressions

Simplify.

1. $\frac{x+2}{x-3} \cdot \frac{x^2-8x+15}{5x-25} = \mathbf{\frac{(x+2)}{5}}$
2. $\frac{12x^2y^4}{36ab^3} \cdot \frac{6a^2v^3}{48xy^4} = \mathbf{\frac{xav^3}{24b^3}}$
3. $\frac{x^2+6x+8}{x^2-16} \cdot \frac{3x-12}{4x+4} = \mathbf{\frac{3(x+2)}{4(x+1)}}$
4. $\frac{x^2-2x-8}{x^2-4} \cdot \frac{x-2}{x+3} = \mathbf{\frac{(x-4)}{(x+3)}}$
5. $\frac{x^2+3x+2}{x+7} \cdot \frac{x^2+9x+14}{x^2+4x+4} = \mathbf{x+1}$
6. $\frac{2x^2+12x+18}{x^2+5x-6} \div \frac{2x+6}{x-1} = \mathbf{\frac{(x+3)}{(x+6)}}$
7. $\frac{2x^2+6x}{x^2+2x} \div \frac{x^2-9}{4x-12} = \mathbf{\frac{8}{(x+2)}}$
8. $\frac{15x^4y^2}{5xy} \div \frac{10x^3y}{5y^2} = \mathbf{\frac{3y^2}{2}}$
9. $\frac{x^2+8x}{x^2+14x+48} \div \frac{x^2+x}{x^2} = \mathbf{\frac{x^2}{(x+6)(x+1)}}$
10. $\frac{3x^2+6x}{x^2+6x} \div \frac{x^2-4}{2x-4} = \mathbf{\frac{6}{(x+6)}}$
11. $\frac{x}{x-4} + \frac{4}{x^2-x-12} = \mathbf{\frac{x^2+3x+4}{[(x-4)(x+3)]}}$
12. $\frac{4}{3x-8} - \frac{x}{4x-7} = \mathbf{\frac{-3x^2+24x-28}{[(4x-7)(3x-8)]}}$
13. $-\frac{4}{2x^2} + \frac{5}{2x^2} + \frac{8}{3x^2} = \mathbf{\frac{19}{6x^2}}$
14. $\frac{4x}{x^2+x-2} - \frac{4x}{x^2+x-2} = \mathbf{0}$

Name _____ HSA-APR.D.7

Operations with Rational Expressions

Simplify.

1. $\frac{2x^2-32}{2x+8} \cdot \frac{x^2-9}{x^2-3x-4} = \mathbf{\frac{(x^2-9)}{(x+1)}}$
2. $\frac{14x^2y^4}{42a^2b^4} \cdot \frac{28a^2b^3}{35x^3y^4} = \mathbf{\frac{4}{15bx}}$
3. $\frac{x^2-100}{x-5} \cdot \frac{x+5}{x^2-5x-50} = \mathbf{\frac{(x+10)}{(x-5)}}$
4. $\frac{x^2-12x+35}{x^2-5x-14} \cdot \frac{x^2+7x+10}{x^2-25} = \mathbf{1}$
5. $\frac{9x^2-25}{4x^2+4x-3} \cdot \frac{40x^2-10x}{3x+5} = \mathbf{\frac{10x(12x^2-24x+5)}{(4x^2+4x-3)}}$
6. $\frac{x+4}{6x^2+24} \cdot \frac{2x^3-8x}{x^2+4x} = \mathbf{\frac{1}{3}}$
7. $\frac{x^2-7x}{x^2-14x+49} \div \frac{2x^2+6x}{x^2+x-56} = \mathbf{\frac{(x+8)}{2(x+3)}}$
8. $\frac{x^2-16}{x^2+7x+12} \div \frac{5x-20}{x+3} = \mathbf{\frac{1}{5}}$
9. $\frac{27x^4y^2}{9x^3y} \div \frac{4x^5y^3}{16x^4y^2} = \mathbf{12x^2}$
10. $\frac{x^2+9x-10}{x^2+5x-14} \div \frac{3x+30}{2x-4} = \mathbf{\frac{(2x-2)}{(3x+21)}}$
11. $\frac{24a^4b^2}{8a^2b} \div \frac{12ab^3}{16a^3b} = \mathbf{\frac{4a^2}{b}}$
12. $\frac{11x}{x^2-6x-7} + \frac{5x}{x^2+9x+8} = \mathbf{\frac{(16x^2+53x)}{[(x-7)(x+1)(x+8)]}}$
13. $\frac{3x}{x^2-4} + \frac{5x}{x^2-4} = \mathbf{\frac{8x}{(x+2)(x-2)}}$
14. $-\frac{10}{4x^2} + \frac{6}{4x^2} + \frac{8}{4x^2} = \mathbf{\frac{1}{x^2}}$
15. $\frac{5}{2xy} + \frac{5}{4xy} - \frac{12}{6xy} = \mathbf{\frac{7}{4xy}}$
16. $\frac{8}{4x+16} - \frac{1}{x+4} - \frac{3}{4} = \mathbf{\frac{-3x-8}{4(x+4)}}$

Name _____ 7.RP.A.1

Ratios and Proportions

Solve the following ratio for x.

$\frac{x}{5} = \frac{4}{10}$ → Take cross products and solve. → $\frac{x}{5} \times \frac{4}{10}$ $5 \cdot 4 = 20$, $x \cdot 10 = 10x$

→ $10x = 20$ → $\frac{10x}{10} = \frac{20}{10}$ → $x = 2$

Solve.

1. $\frac{4}{(x-3)} = \frac{28}{49}$ $\mathbf{x = 10}$
2. $\frac{(5+x)}{10} = \frac{2}{5}$ $\mathbf{x = -1}$
3. $\frac{x}{30} = \frac{7}{10}$ $\mathbf{x = 21}$
4. $\frac{(x-2)}{16} = \frac{x}{4}$ $\mathbf{x = -\frac{2}{3}}$
5. $\frac{2}{x} = \frac{6}{30}$ $\mathbf{x = 10}$
6. $\frac{(x+1)}{7} = \frac{6}{14}$ $\mathbf{x = 2}$
7. $\frac{x}{15} = \frac{5}{75}$ $\mathbf{x = 1}$
8. $\frac{x}{20} = \frac{2}{10}$ $\mathbf{x = 4}$
9. $\frac{x}{6} = \frac{(x-3)}{12}$ $\mathbf{x = -3}$
10. $\frac{x}{5} = \frac{12}{6}$ $\mathbf{x = 10}$
11. $\frac{6}{(x+5)} = \frac{18}{24}$ $\mathbf{x = 3}$
12. $\frac{5}{15} = \frac{x}{9}$ $\mathbf{x = 3}$
13. $\frac{+10}{} = \frac{5}{2}$ $\mathbf{x = \frac{25}{2}}$
14. $\frac{x}{3} = \frac{12}{27}$ $\mathbf{x = \frac{4}{3}}$

Answer Key

Name _____ (6.RP.A.3, 7.RP.A.1, 7.RP.A.2c, 7.RP.A.3)

Ratios and Proportions

Three liters of soda cost $3.00. At this rate, how much would 10 liters of soda cost? To find the cost, write and solve a ratio using x to represent the cost.

$\frac{\text{liters}}{\text{cost}} \rightarrow \frac{3}{3.00} = \frac{10}{x} \rightarrow 3x = 10(3.00) \rightarrow 3x = 30 \rightarrow \frac{3x}{3} = \frac{30}{3} \rightarrow x = 10 \rightarrow$ The cost of 10 liters of soda is $10.00.

Solve.

1. $\frac{4}{x+8} = \frac{3}{x}$ **$x = 24$**
2. $\frac{x}{4} = \frac{21}{7}$ **$x = 12$**
3. $\frac{9}{3} = \frac{6}{x}$ **$x = 2$**
4. $\frac{x-3}{4} = \frac{x-4}{3}$ **$x = 7$**
5. $\frac{x}{4} = \frac{8}{2x}$ **$x = 4$**
6. $\frac{12}{x} = 3$ **$x = 4$**

Set up a proportion and solve.

7. A copy machine can print 120 pages per minute. At this rate, how many minutes will it take to make 840 copies?

$\frac{\text{minutes}}{\text{copies}} = \frac{1}{120} = \frac{x}{840}$ **$x = 7$ minutes**

8. Two gallons of fruit juice will serve 35 people. How much fruit juice is needed to serve 105 people?

$\frac{\text{gallons}}{\text{people}} = \frac{2}{35} = \frac{x}{105}$ **$x = 6$ gallons**

9. A stock investment of $4,000 earns $360 in interest each year. At the same rate, how much interest will a person earn if he invests $6,000?

$\frac{\text{interest}}{\text{investment}} = \frac{\$360}{\$4{,}000} = \frac{x}{\$6{,}000}$ **$x = \$540$**

Name _____ (6.RP.A.3, 7.RP.A.1, 7.RP.A.2c, 7.RP.A.3)

Ratios and Proportions

Set up a proportion and solve.

1. An investment of $36,000 earns $900 each year. At the same rate, how much money must be invested to earn $1,200 each year?

$\frac{\text{investment}}{\text{earning}} = \frac{\$36{,}000}{\$900} = \frac{x}{\$1{,}200}$ **$x = \$48{,}000$**

2. The sales tax on a $15,000 car is $540. At this rate, what is the tax on a $32,000 car?

$\frac{\text{tax}}{\text{value}} = \frac{\$540}{\$15{,}000} = \frac{x}{\$32{,}000}$ **$x = \$1{,}152$**

3. Six gallons of paint will cover 120 doors. At this rate, how many gallons of paint are needed to cover 480 doors?

$\frac{\text{gallons}}{\text{doors}} = \frac{6}{120} = \frac{x}{480}$ **$x = 24$ gallons**

4. A lawnmower can cut 1 acre on 0.5 gallons of gasoline. At this rate, how much gasoline is needed to cut 3.5 acres?

$\frac{\text{gallons}}{\text{acres}} = \frac{0.5}{1} = \frac{x}{3.5}$ **$x = 1.75$ gallons**

5. An aerobics instructor burns 400 calories in 1 hour. How many hours would the instructor have to do aerobics to burn 660 calories?

$\frac{\text{hours}}{\text{calories}} = \frac{1}{400} = \frac{x}{660}$ **$x = 1.65$ hours**

6. The real estate tax for a house that costs $56,000 is $1,400. At this rate, what is the value of a house for which the real estate tax is $1,800?

$\frac{\text{value}}{\text{tax}} = \frac{\$56{,}000}{\$1{,}400} = \frac{x}{\$1{,}800}$ **$x = \$72{,}000$**

7. One hundred thirty-six tiles are required to tile a 36 ft.^{2} area. At this rate, how many tiles are required to tile a 288 ft.^{2} area?

$\frac{\text{tiles}}{\text{area}} = \frac{136}{36} = \frac{x}{288}$ **$x = 1{,}088$ tiles**

Name _____ (5.G.A.1, 6.NS.C.6b, 6.NS.C.6c)

Graphing Ordered Pairs

Start at the origin (0, 0).
(x, y) = (1, −2) right 1 and down 2
(x, y) = (−3, 4) left 3 and up 4
(x, y) = (2, −2) right 2 and down 2

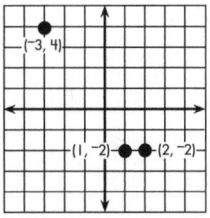

Label the following points. Start at the origin (0, 0).

A (6, 2)
B (6, 3)
C (5, 1)
D (5, 0)
E (1, −3)
F (2, −6)
G (5, 4)
H (2, 4)
I (4, 3)
J (1, −2)
K (5, −3)
L (1, 4)
M (3, 1)

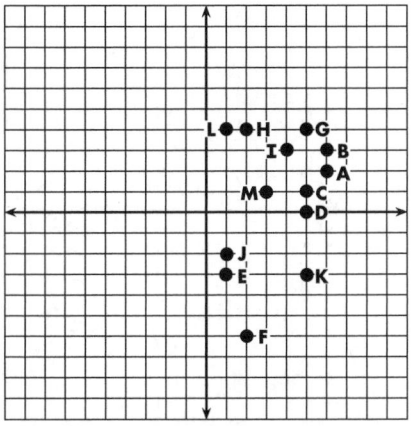

Name _____ (5.G.A.1, 6.NS.C.6b, 6.NS.C.6c)

Graphing Ordered Pairs

Graph and label the following points.

A (−1, 2) B (−1, 0) C (−1, −6) D (−4, 2)
E (−6, 2) F (−5, 6) G (−4, 1) H (2, −8)
I (6, 6) J (−7, 5) K (−1, −1) L (−3, −3)
M (6, −4)

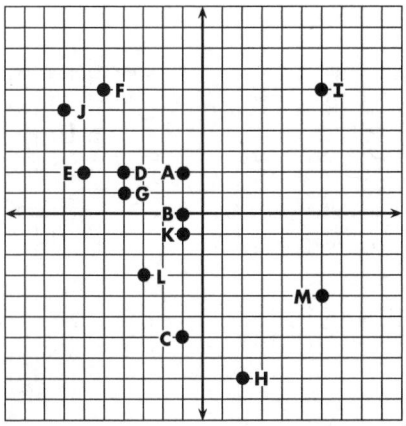

Answer Key

Name _____ 6.NS.C.6, 6.EE.B.6, 8.F.A.1, HSA-REI.D.10

Graphing Ordered Pairs

Solve for each x value. Graph each ordered pair.

1. $x + y = 4$

x	y
4	**0**
5	**−1**
6	**−2**

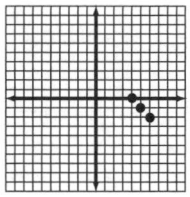

2. $y = 3 - x$

x	y
5	**−2**
−2	**5**
0	**3**

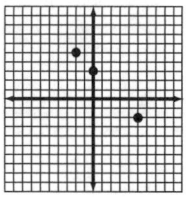

3. $4x + y = 6$

x	y
0	**6**
1	**2**
2	**−2**

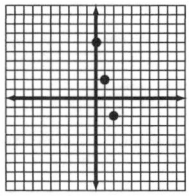

4. $y = 2x - 7$

x	y
3	**−1**
4	**1**
5	**3**

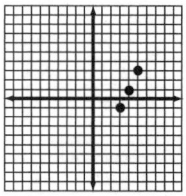

5. $y = x + 3$

x	y
2	**5**
0	**3**
−3	**0**

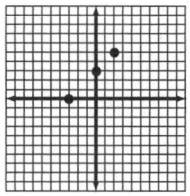

6. $y = 9 - x$

x	y
0	**9**
4	**5**
3	**6**

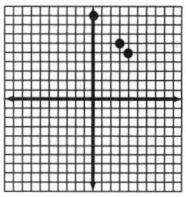

Name _____ 8.EE.B.6, 8.F.A.1

Graphing Linear Equations

One way to graph a linear equation is to find the x- and y-intercepts. To find the x-intercept, let $y = 0$. To find the y-intercept, let $x = 0$. Plot the two points and connect with a straight line.

Graph $2x - y = 2$ by using the x- and y-intercepts.

x-intercept	**y-intercept**
$2x - y = 2$	$2x - y = 2$
$2x - 0 = 2$	$2(0) - y = 2$
$2x = 2$	$-y = 2$
$x = 1$	$y = -2$
$(1, 0)$	$(0, -2)$

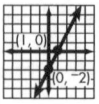

Find the x- and y-intercepts and graph.

1. $3x - 9y = 18$

(0, −2)
(6, 0)

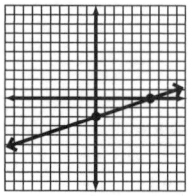

2. $4x + 2y = -8$

(0, −4)
(−2, 0)

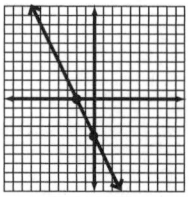

3. $2x + 4y = 20$

(0, 5)
(10, 0)

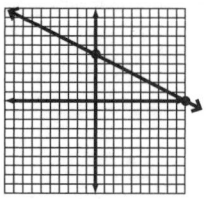

4. $x + 5y = 10$

(0, 2)
(10, 0)

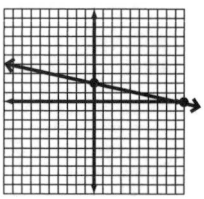

Name _____ 8.EE.B.6, 8.F.A.1

Graphing Linear Equations

Another way to graph a linear equation is to use the **slope-intercept form**. A linear equation written in slope-intercept form is $y = mx + b$, where m is the slope of the line and b is the y-intercept.

Example: $y = 3x + 2$ — $m = 3$, y-int. $= 2$; $y = -5x + 1$ — $m = -5$, y-int. $= 1$; $y = x - 7$ — $m = 1$, y-int. $= -7$

You can rewrite an equation in slope-intercept form by adding or subtracting terms from both sides and simplifying.

To graph the equation, plot the y-intercept. Starting at the y-intercept, move up or down and then right or left according to the slope and make a new point.

Example: If $m = 4$ and $b = 3$, the y-intercept is (0, 3). The next point to graph would be (1, 7) since the slope is 4, or $\frac{4}{1}$.

Solve for y, state the m and b, and graph.

1. $y = -3x + 5$

$m = -3$
$b = 5$

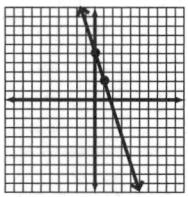

2. $-8x + y = -4$

$m = 8$
$b = -4$

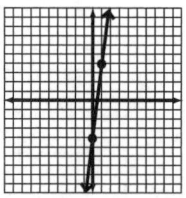

3. $x - y = 7$

$m = 1$
$b = -7$

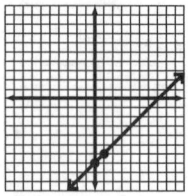

4. $2y = 3x - 4$

$m = \frac{3}{2}$
$b = -2$

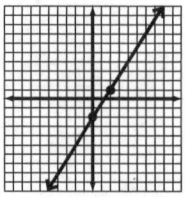

Name _____ 8.EE.B.6, 8.F.A.1

Graphing Linear Equations

Graph each equation using either slope-intercept form or x- and y-intercepts.

1. $2x + 3y = 9$

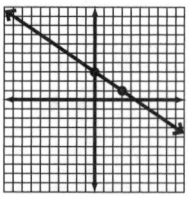

2. $x - 3y + 9 = 0$

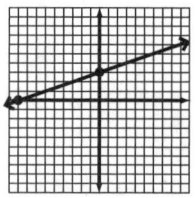

3. $6x + 2y = 12$

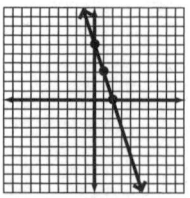

4. $y = 2x - 4$

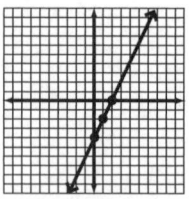

5. $5x - y = 7$

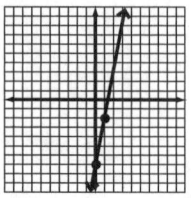

6. $x - 3y + 6 = 0$

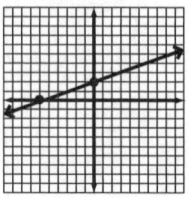

Answer Key

Name _____ (8.EE.B.6, 8.F.A.3)

Writing Linear Equations

This equation, $y = mx + b$, is the **slope-intercept of a straight line.**
For all equations of the form $y = mx + b$, m is the slope of the line.
The y-intercept is $(0, b)$.

Find the equation of a line using the slope-intercept form: $y = mx + b$.

1. $m = 4$, y-int. $= -1$
   $y = 4x - 1$
2. $m = -2$, y-int. $= -5$
   $y = -2x - 5$
3. $m = 1$, y-int. $= 2$
   $y = x + 2$
4. $m = \frac{7}{5}$ $b = -2$
   $y = \frac{7}{5}x - 2$
5. $m = \frac{3}{4}$ $b = \frac{2}{3}$
   $y = \frac{3}{4}x + \frac{2}{3}$
6. $m = \frac{3}{7}$ $b = \frac{1}{3}$
   $y = \frac{3}{7}x + \frac{1}{3}$
7. $m = \frac{3}{5}$ $b = \frac{1}{5}$
   $y = \frac{3}{5}x + \frac{1}{5}$
8. $m = -4$ $b = \frac{3}{4}$
   $y = -4x + \frac{3}{4}$
9. $m = 0$ $b = 5$
   $y = 5$
10. $m = \frac{1}{4}$ $b = \frac{2}{3}$
    $y = \frac{1}{4}x + \frac{2}{3}$

Name _____ (8.EE.B.6, 8.F.A.3)

Writing Linear Equations

Slope-Intercept Formula
$y = mx + b$
m: slope
b: y-intercept containing coordinate points (o, b)

$m = 4$, passing through points $(1, 2)$
Using this information, $(m = 4, x = 1, y = 2)$.
Substitute into $y = mx + b$ to find b.
$2 = (4)(1) + b \longrightarrow 2 = 4 + b$
$\longrightarrow 2 - 4 = 4 - 4 + b \longrightarrow -2 = b$
With the information $m = 4$, $b - -2$, write the equation as $y = 4x - 2$.

Find the equation of the line with the given slope passing through the indicated point (P).

1. $p = (-2, -6)$, $m = 3$
   $b =$ **0**
   Equation **$y = 3x$**
2. $p = (-4, 3)$, $m = 1$
   $b =$ **7**
   Equation **$y = x + 7$**
3. $p = (3, 5)$, $m = 0$
   $b =$ **5**
   Equation **$y = 5$**
4. $p = (5, 7)$, $m = 3$
   $b =$ **-8**
   Equation **$y = 3x - 8$**
5. $p = (-7, -7)$, $m = -7$
   $b =$ **-56**
   Equation **$y = -7x - 56$**
6. $p = (2, -6)$, $m = 4$
   $b =$ **-14**
   Equation **$y = 4x - 14$**
7. $p = (2, 4)$, $m = 4$
   $b =$ **-4**
   Equation **$y = 4x - 4$**
8. $p = (6, -1)$, $m = -5$
   $b =$ **29**
   Equation **$y = -5x + 29$**
9. $p = (-1, 1)$, $m = 2$
   $b =$ **3**
   Equation **$y = 2x + 3$**
10. $p = (-1, -6)$, $m = 2$
    $b =$ **-4**
    Equation **$y = 2x - 4$**
11. $p = (4, 5)$, $m = -2$
    $b =$ **13**
    Equation **$y = -2x + 13$**
12. $p = (2, 6)$, $m = -6$
    $b =$ **18**
    Equation **$y = -6x + 18$**

Name _____ (8.EE.B.6, 8.F.A.3)

Writing Linear Equations

Write the equation of the line that passes through the point and has the given slope. Put each equation in slope-intercept form.

1. $p = (2, 3)$, $m = 4$
   $y = 4x - 5$
2. $p = (-5, -4)$, $m = -8$
   $y = -8x - 44$
3. $p = (-5, 7)$, $m = -3$
   $y = -3x - 8$
4. $p = (2, 4)$, $m = -1$
   $y = -x + 6$
5. $p = (-1, 2)$, $m = 1$
   $y = x + 3$
6. $p = (6, -8)$, $m = 2$
   $y = 2x - 20$
7. $p = (1, -1)$, $m = \frac{1}{2}$
   $y = \frac{1}{2}x - \frac{3}{2}$
8. $p = (-4, -1)$, $m = -\frac{1}{2}$
   $y = -\frac{1}{2}x - 3$
9. $p = (6, 9)$ $m = 5$
   $y = 5x - 21$
10. $p = (-1, 2)$ $m = -1$
    $y = -x + 1$
11. $p = (-2, 9)$ $m = 3$
    $y = 3x + 15$
12. $p = (-2, -1)$ $m = 2$
    $y = 2x + 3$
13. $p = (3, 10)$ $m = -5$
    $y = -5x + 25$
14. $p = (7, 8)$ $m = 4$
    $y = 4x - 20$

Name _____ (HSA-REI.D.12)

Graphing Linear Inequalities

Graph the line $y > 2x + 3$.

1. $m = \frac{2}{1}$ $b = 3$
2. If $>$ or $<$, connect the points with a dotted line.
3. If $\geq$ or $\leq$, connect the points with a solid line.
   The coordinate plane is now divided into 2 regions.
4. Test any (x, y) on each side of the line in the original inequality.

$y > 2x + 3$
Test point A $(-1, 4)$.
Is $4 > 2(-1) + 3$?
$4 > -2 + 3$
$4 > 1 \longrightarrow$ true
(Shade this region)
Test point A.

$y > 2x + 3$
Test point B $(0, 0)$.
Is $0 > (0) + 3$?
$0 > 0 + 3$
$0 > 3 \longrightarrow$ false
(Do not shade this region)
Test point B.

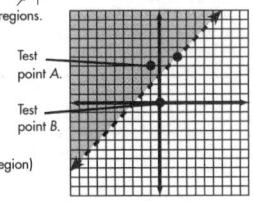

Graph the solution set.

1. $x + 4y > 8$ **$y > -\frac{1}{4}x + 2$**

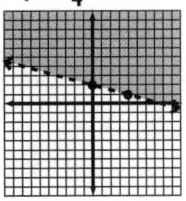

2. $-2x + 2y \geq 10$ **$y \geq x + 5$**

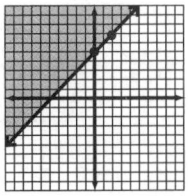

Answer Key

Name _____ HSA-REI.D.12

Graphing Linear Inequalities

Graph the solution set.

1. $3x + 2y \geq 6$ **$y \geq -\frac{3}{2}x + 3$**

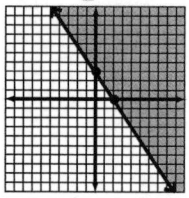

2. $2x + y < 4$ **$y < -2x + 4$**

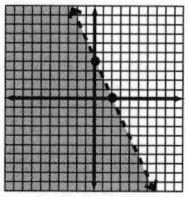

3. $6x - 3y > 15$ **$y < 2x - 5$**

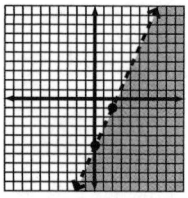

4. $2x + 3y \geq 6$ **$y \geq -\frac{2}{3}x + 2$**

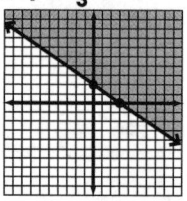

5. $3x - 4y \leq 12$ **$y \geq \frac{3}{4}x - 3$**

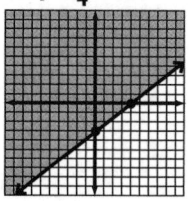

6. $4x + 2y < 6$ **$y < -2x + 3$**

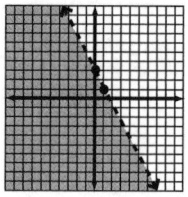

Name _____ HSA-REI.D.12

Graphing Linear Inequalities

Graph the solution set.

1. $x + 2y < 0$ **$y < -\frac{x}{2}$**

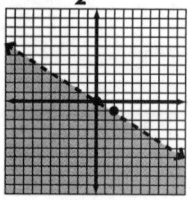

2. $5x - 2y \leq 10$ **$y \geq \frac{5}{2}x - 5$**

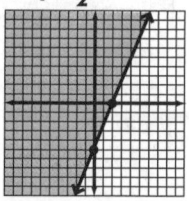

3. $6x - 3y > 18$ **$y < 2x - 6$**

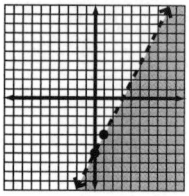

4. $2x - 5y < 10$ **$y > \frac{2}{5}x - 2$**

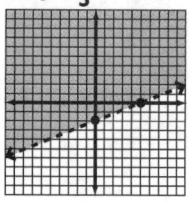

5. $-5x + 5y \leq 15$ **$y \leq x + 3$**

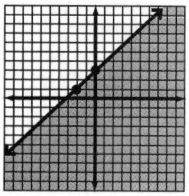

6. $y + 6 > 0$ **$y > -6$**

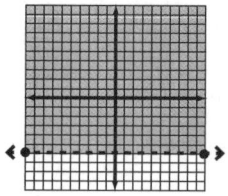

Name _____ 8.EE.C.8b, HSA-REI.C.6, HSA-REI.D.11

Solving Systems of Equations by Graphing

When two or more equations are considered together, it is called a **system of equations**. The following example is a system of two linear equations in two variables.

$x + 2y = 4$
$2x + y = -1$

The graphs of these equations are straight lines.

An ordered pair that is a solution of each equation of the system is a solution of the system of equations in two variables. The solution of a system of linear equations can be found by graphing the lines of the system. The solution of the system of equations is the point where the lines intersect.

Solve by graphing:
$x + 2y = 4$
$2x + y = -1$

Graph each line and find the point of intersection.

The solution is $(-2, 3)$ because the ordered pair lies on each line.

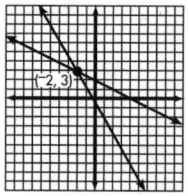

Solve by graphing.

1. $x + y = 7$
$3x - y = -3$

(1, 6)

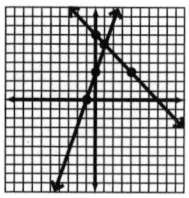

2. $x + y = 4$
$x - y = 6$

(5, −1)

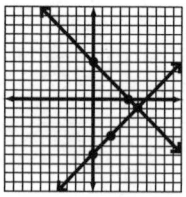

Name _____ 8.EE.C.8b, HSA-REI.C.6, HSA-REI.D.11

Solving Systems of Equations by Graphing

Solve by graphing.

1. $3x - 4y = 12$
$2x + 4y = -12$

(0, −3)

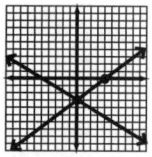

2. $4x - 2y = 8$
$y = 2$

(3, 2)

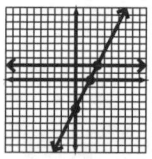

3. $x - y = 1$
$x + 2y = 10$

(4, 3)

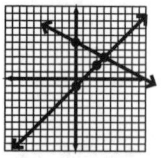

4. $x = 5$
$y = -1$

(5, −1)

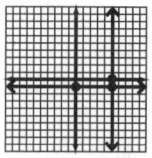

5. $y = 2x + 4$
$y = x + 6$

(2, 8)

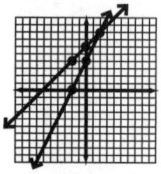

6. $x - y = -4$
$3x - y = -12$

(−4, 0)

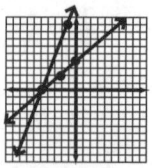

7. $x + y = 3$
$x - y = 5$

(4, −1)

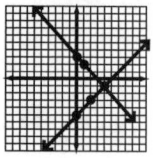

8. $x = 4$
$6x - 2y = 4$

(4, 10)

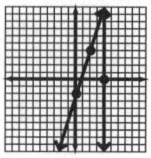

© Carson-Dellosa • CD-104632

Answer Key

Name _____ (8.EE.C.8b, HSA-REI.C.6, HSA-REI.D.11)

Solving Systems of Equations by Graphing

Solve by graphing.

1. $y = x - 3$
   $x = 2y + 6$
   (0, −3)

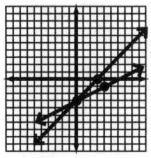

2. $-2x + y = 0$
   $6x + y = -16$
   (−2, −4)

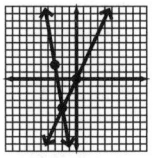

3. $y = 3x + 5$
   $y = 1 - x$
   (−1, 2)

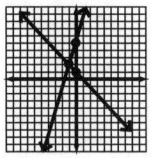

4. $y = x - 5$
   $y + 2x = 1$
   (2, −3)

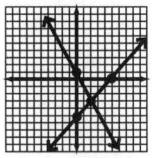

5. $x - 4y = 2$
   $x + 2y = 8$
   (6, 1)

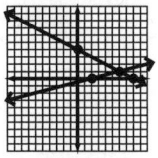

6. $x + y = 6$
   $3x - y = 2$
   (2, 4)

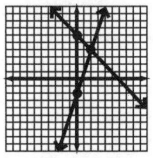

7. $2x + y = 6$
   $3x + y = 12$
   (6, −6)

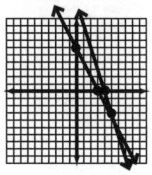

8. $y = x + 3$
   $2y = 3x + 1$
   (5, 8)

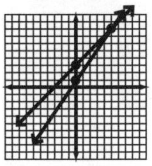

Name _____ (8.EE.C.8b, HSA-REI.C.6)

Solving Systems of Equations

$$\begin{array}{r} 2x + 3y = 7 \\ +\ \ x - 3y = -1 \\ \hline 3x \qquad = 6 \\ \frac{3x}{3} = \frac{6}{3} \\ \mathbf{x = 2} \end{array} \rightarrow \begin{array}{r} 2x + 3y = 7 \\ 2(2) + 3y = 7 \\ 4 + 3y = 7 \\ 3y = 3 \\ \mathbf{y = 1} \end{array}$$

Substitute 2 for x in $2x + 3y = 7$. answer (2, 1)

Solve.

1. $2x - 5 = y$
   $x - 7 = -y$
   (4, 3)

2. $x + 4y = 2$
   $-x + y = 8$
   (−6, 2)

3. $y = 2x - 2$
   $-y = x$
   $(\frac{2}{3}, -\frac{2}{3})$

4. $3x + y = 8$
   $3x - y = 4$
   (2, 2)

5. $2x - y = 6$
   $3x + y = 4$
   (2, −2)

6. $y = 5x + 1$
   $2y = -5x + 2$
   (0, 1)

7. $x + y = 7$
   $x - y = 3$
   (5, 2)

8. $3x + y = 5$
   $x - y = 7$
   (3, −4)

9. $3x - 4y = 14$
   $x + 4y = 2$
   $(4, -\frac{1}{2})$

10. $5x - 3y = -1$
    $4x + 3y = 10$
    (1, 2)

11. $8x - 3y = 1$
    $-8x + 5y = 9$
    (2, 5)

12. $3y - 4x = 5$
    $y + 4x = 7$
    (1, 3)

Name _____ (8.EE.C.8b, HSA-REI.C.6)

Solving Systems of Equations

1. Solve.
   $6x + 5y = 6$
   $6x - 3y = 6$

2. $\left.\begin{array}{r} 6x + 5y = 6 \\ (-1)(6x - 3y = 6) \end{array}\right\}$ Multiply to create the additive inverse.

3. $\left.\begin{array}{r} 6x + 5y = 6 \\ -6x + 3y = -6 \\ \hline 8y = 0 \\ y = 0 \end{array}\right\}$ Use the addition method.

4. $\left.\begin{array}{r} 6x + 5y = 6 \\ 6x + 5(0) = 6 \\ 6x = 6 \\ x = 1 \end{array}\right\}$ Substitute.

Solve.

1. $3x + 8y = 8$
   $2x + y = 1$
   (0, 1)

2. $3x - 4y = 4$
   $x - y = 1$
   (0, −1)

3. $4x - 4y = 16$
   $2x + 2y = 4$
   (3, −1)

4. $2x - 5y = 21$
   $x + y = 7$
   (8, −1)

5. $4x + 8y = 20$
   $x - y = 2$
   (3, 1)

6. $6x + 4y = -22$
   $3x + 6y = -7$
   $(-\frac{13}{3}, 1)$

7. $2x + y = 3$
   $x + 3y = 4$
   (1, 1)

8. $3x + 3y = -3$
   $x - y = -5$
   (−3, 2)

9. $4x + 2y = 10$
   $x + 2y = 1$
   (3, −1)

10. $3x + 5y = -8$
    $x + 7y = -8$
    (−1, −1)

11. $x - y = 7$
    $x + 2y = 1$
    (5, −2)

12. $2x - 8y = 6$
    $-x + 7y = 9$
    (−5, −2)

Name _____ (8.EE.C.8b, HSA-REI.C.6)

Solving Systems of Equations

Solve.

1. $y = 3 - 2x$
   $y = 2 - 3x$
   (−1, 5)

2. $x + y = 5$
   $x = y + 7$
   (6, −1)

3. $x - y = 1$
   $2x + y = 8$
   (3, 2)

4. $3x - y = 9$
   $y = x + 5$
   (7, 12)

5. $3x + 4y = 26$
   $-2x + y = 1$
   (2, 5)

6. $y = 2x + 3$
   $y = 4x + 4$
   $(-\frac{1}{2}, 2)$

7. $2x + 7y = 8$
   $x + 5y = 7$
   (−3, 2)

8. $y = 4x + 4$
   $y = 2x + 8$
   (2, 12)

9. $x + 3y = 17$
   $2x + 3y = 22$
   (5, 4)

10. $4x - 7y = 9$
    $y = x - 3$
    (4, 1)

11. $8x - 5y = 9$
    $y = 2x - 4$
    $(\frac{11}{2}, 7)$

12. $2x + 4y = -2$
    $3x + y = 7$
    (3, −2)

13. $3x + y = 5$
    $2x + 3y = 8$
    (1, 2)

14. $2x + 6y = 24$
    $x - 4y = -2$
    (6, 2)

Answer Key

Name _____ HSN-RN.A.2

Operations With Radicals

Simplify: $\sqrt{49x^4y^8} = 7\sqrt{x^4y^8} = 7x^2y^4$

Multiply: $\sqrt{3a} \cdot \sqrt{4a} = \sqrt{12a^2} = \sqrt{3} \cdot 4 \cdot a^2 = 2a\sqrt{3}$

Divide: $\sqrt{\frac{18}{2}} = \sqrt{9} = 3$ $\quad \sqrt{\frac{9}{25}} = \frac{\sqrt{9}}{\sqrt{25}} = \frac{3}{5}$

Add or subtract: $2\sqrt{y} + 3\sqrt{y} + \sqrt{y} = 6\sqrt{y}$ $\quad \sqrt{4x} + 3\sqrt{x} = 2\sqrt{x} + 3\sqrt{x} = 3\sqrt{x}$

Simplify.

1. $\sqrt{x^2y^{10}} =$ $\mathbf{xy^5}$
2. $\sqrt{27x^8} =$ $\mathbf{3x^4\sqrt{3}}$
3. $\sqrt{x^{16}} =$ $\mathbf{x^8}$
4. $\sqrt{125b^{15}} =$ $\mathbf{3b^7\sqrt{5b}}$
5. $\sqrt{x^2y^4} \cdot 2\sqrt{xy} =$ $\mathbf{2xy^2\sqrt{xy}}$
6. $\sqrt{9} \cdot \sqrt{32} =$ $\mathbf{12\sqrt{2}}$
7. $3\sqrt{5} \cdot 2\sqrt{4} =$ $\mathbf{12\sqrt{5}}$
8. $2\sqrt{4x^3y} \cdot y\sqrt{x^5y^7} =$ $\mathbf{4x^4y^5}$
9. $\sqrt{\frac{36}{9}} =$ $\mathbf{2}$
10. $\sqrt{\frac{27x}{3x}} =$ $\mathbf{3}$
11. $3\sqrt{x^3} - 4\sqrt{x^3} =$ $\mathbf{-x\sqrt{x}}$
12. $3\sqrt{2y} + 2\sqrt{2y} =$ $\mathbf{5\sqrt{2y}}$
13. $2\sqrt{y} - 4\sqrt{y} =$ $\mathbf{-2\sqrt{y}}$
14. $3y\sqrt{2y} - y\sqrt{2y} =$ $\mathbf{2y\sqrt{2y}}$

© Carson-Dellosa • CD-104632

Name _____ HSN-RN.A.2

Operations With Radicals

Simplify.

1. $\sqrt{x^{14}y^6} =$ $\mathbf{x^7y^3}$
2. $\sqrt{169y^{12}} =$ $\mathbf{13y^6}$
3. $\sqrt{16x^4} =$ $\mathbf{4x^2}$
4. $\sqrt{8x^3} =$ $\mathbf{2x\sqrt{2x}}$
5. $\sqrt{81x^6} =$ $\mathbf{9x^3}$
6. $\sqrt{25x^6} =$ $\mathbf{5x^3}$
7. $4\sqrt{9x^3} \cdot 3\sqrt{4x} =$ $\mathbf{72x^2}$
8. $x\sqrt{5x^3y} \cdot x\sqrt{5x^3y} =$ $\mathbf{5x^4y\sqrt{x}}$
9. $2\sqrt{9x^2} \cdot 2\sqrt{4x^2} =$ $\mathbf{24x^2}$
10. $6\sqrt{9xy} \cdot 4\sqrt{2xy} =$ $\mathbf{72xy\sqrt{2}}$
11. $2\sqrt{4x^3y} \cdot 3\sqrt{3a^2b^2} =$ $\mathbf{12abx\sqrt{3xy}}$
12. $\sqrt{\frac{x^2}{25}} =$ $\mathbf{\frac{x}{5}}$
13. $\sqrt{\frac{8x^3}{2x}} =$ $\mathbf{2x}$
14. $\sqrt{\frac{18x^3}{2x}} =$ $\mathbf{3x}$
15. $x\sqrt{27} + x\sqrt{12} =$ $\mathbf{5x\sqrt{3}}$
16. $3\sqrt{6x} + 5\sqrt{6x} =$ $\mathbf{8\sqrt{6x}}$
17. $4\sqrt{2x^3} + 3\sqrt{2x^3} =$ $\mathbf{7x\sqrt{2x}}$
18. $2\sqrt{50} - 4\sqrt{8} - 3\sqrt{72} =$ $\mathbf{-16\sqrt{2}}$

© Carson-Dellosa • CD-104632

Name _____ HSN-RN.A.2

Operations with Radicals

Simplify.

1. $\sqrt{x^8y^9} =$ $\mathbf{x^4y^4\sqrt{xy}}$
2. $\sqrt{54x^8} =$ $\mathbf{3x^4\sqrt{6}}$
3. $\sqrt{9a^4b^8} =$ $\mathbf{3a^2b^4}$
4. $\sqrt{49x^4y^2} =$ $\mathbf{7x^2y}$
5. $5\sqrt{2x^6y} \cdot 3\sqrt{3x^3y^5} =$ $\mathbf{15x^4y^3\sqrt{6x}}$
6. $4\sqrt{8a^6b} \cdot 4\sqrt{8a^4b^4} =$ $\mathbf{128a^5b^2\sqrt{b}}$
7. $3\sqrt{2x^3} \cdot 3\sqrt{3x^2y^2} =$ $\mathbf{9x^2y\sqrt{6x}}$
8. $5\sqrt{4a} \cdot 2\sqrt{6a} =$ $\mathbf{20a\sqrt{6}}$
9. $x\sqrt{3x} \cdot \sqrt{3x^3} =$ $\mathbf{3x^3}$
10. $\sqrt{2x^4} \cdot \sqrt{10x^2y^2} =$ $\mathbf{2x^3y\sqrt{5}}$
11. $\sqrt{\frac{9}{64}} =$ $\mathbf{\frac{3}{8}}$
12. $\sqrt{\frac{50x^2}{2}} =$ $\mathbf{5x}$
13. $\sqrt{\frac{49x^2}{25x^3}} =$ $\mathbf{\frac{7}{5\sqrt{x}}}$
14. $\sqrt{\frac{12x^2}{60}} =$ $\mathbf{\frac{x}{\sqrt{5}}}$
15. $\sqrt{\frac{3x^7}{108y^2}} =$ $\mathbf{\frac{x^3\sqrt{x}}{6y}}$
16. $4\sqrt{y^3} - 2\sqrt{y^3} =$ $\mathbf{2y\sqrt{y}}$
17. $y\sqrt{y^4} - 2y\sqrt{y^4} =$ $\mathbf{-y^3}$
18. $4x\sqrt{x^3} + 2x\sqrt{x^3} =$ $\mathbf{6x^2\sqrt{x}}$
19. $4\sqrt{x} - 2\sqrt{x} - 3\sqrt{x} + 5\sqrt{x} =$ $\mathbf{4\sqrt{x}}$
20. $3\sqrt{4x^2y} - 8y\sqrt{y} =$ $\mathbf{(6x - 8y)\sqrt{y}}$

© Carson-Dellosa • CD-104632

© Carson-Dellosa • CD-104632

Congratulations!

_____

receives this award for

_____

Signed _____ **Date** _____

© Carson–Dellosa

$x + 6 = {}^{-}6$	$x - 3 = {}^{-}6$	$3x \div 3 = 12$	$\frac{x}{2} = 24$
$5x - 17 = 8$	$\frac{x}{3} = {}^{-}6$	${}^{-}7x + {}^{-}4 = 17$	$x - 4 = {}^{-}3$
$x + 5 = {}^{-}7$	$x - 4 = 6$	$x + 9 = {}^{-}12$	$4 - x = 6$
${}^{-}4x = 36$	$x + 4 = {}^{-}10$	$x - 5 = {}^{-}3$	$3x - 3 = 12$

© CD

$x = 48$	$x = 12$	$x = -3$	$x = -12$
$x = 1$	$x = -3$	$x = -18$	$x = 5$
$x = -2$	$x = -21$	$x = 10$	$x = -12$
$x = 5$	$x = 2$	$x = -14$	$x = -9$

$(3xy)^{-1}$	$(\frac{3}{4})^{-2}$	$(x^3y^2)^3$	$(\frac{x^3}{y^2})^4$
$(3x^2)(4x^3)$	$\frac{(x^6 \cdot x^4)}{x^2}$	$\frac{x^{12}}{x^3}$	$\frac{18x^3}{3x}$
$(2x^3)^3$	$(6x^{-2})^2$	$7x^{-6}$	$x^3 \cdot x \cdot x \cdot x^2$
$(3x^2y^2)^3$	$(-7x^3)^{-2}$	$(-3)^{-2}$	$x^4 \cdot x^4 \cdot x^4$

© CD

$\frac{x^{12}}{y^8}$	x^9y^6	$\frac{16}{9}$	$\frac{1}{3xy}$
$6x^2$	x^9	x^8	$12x^5$
x^7	$\frac{7}{x^6}$	$\frac{36}{x^4}$	$8x^9$
x^{12}	$\frac{1}{9}$	$\frac{1}{49x^6}$	$27x^6y^6$

2^{-3}	$x^2 \cdot x^4$	$\frac{x^6}{x^3}$	$(-2)^3$
$5x^{-3}$	$(\frac{1}{2})^{-3}$	$(x^3)^3$	$x^3 \cdot x^2 \cdot x^3$
$(x^6)^3$	$x^3 \cdot x^3$	$(x^8)^2$	$(5x^3)^2$
$x^6 \cdot x^{-3}$	$3x^{-3}$	$(4x)^{-2}$	$x^3 \cdot x^{-7}$

© CD

-8	x^3	x^6	$\frac{1}{8}$
x^8	x^9	8	$\frac{5}{x^3}$
$25x^6$	x^{16}	x^6	x^{18}
$\frac{1}{x^4}$	$\frac{1}{16x^2}$	$\frac{3}{x^3}$	x^3

Factor.

$6x + 12y$

© CD

Factor.

$3x + 4x^2$

© CD

Factor.

$8 - 2y$

© CD

Factor.

$36x^2 + 6x + 12$

© CD

Factor.

$4x^2 + 64$

© CD

Factor.

$x^2 - 4$

© CD

Factor.

$9x^2 - 16$

© CD

Factor.

$81 - x^2$

© CD

Factor.

$x^2 - 9$

© CD

Factor.

$36x^2 - 81$

© CD

Factor.

$36x^2 - 49y^2$

© CD

Factor.

$25x^2 - y^2$

© CD

Factor.

$3x^2 - 36$

© CD

Factor.

$5x^2 - 35$

© CD

Factor.

$x^2 + 5x + 6$

© CD

Factor.

$x^2 + 6x + 9$

© CD

$6(6x^2 + x + 2)$ $2(4 - y)$ $x(3 + 4x)$ $6(x + 2y)$

$(9 - x)(9 + x)$ $(3x - 4)(3x + 4)$ $(x - 2)(x + 2)$ $4(x^2 + 16)$

$(5x - y)(5x + y)$ $(6x + 7y)(6x - 7y)$ $9(2x - 3)(2x + 3)$ $(x - 3)(x + 3)$

$(x + 3)(x + 3)$ $(x + 3)(x + 2)$ $5(x^2 - 7)$ $3(x^2 - 12)$

Factor.

$x^2 + 9x + 14$

© CD

Factor.

$x^2 + 10x + 16$

© CD

Factor.

$x^2 - 8x + 15$

© CD

Factor.

$x^2 - 12x + 36$

© CD

Factor.

$x^2 - 10x + 24$

© CD

Factor.

$x^2 + 11x + 28$

© CD

Factor.

$x^2 - 5x - 6$

© CD

Factor.

$x^2 - x - 72$

© CD

Factor.

$x^2 - 3x - 10$

© CD

Factor.

$x^2 - 2x - 8$

© CD

Factor.

$x^2 + 4x - 21$

© CD

Factor.

$x^2 + 2x - 15$

© CD

Factor.

$x^2 + 2x - 24$

© CD

Factor.

$x^2 - 13x - 48$

© CD

Factor.

$3x^4 - 243$

© CD

Factor.

$2x^2 - 4x - 48$

© CD

$(x - 6)(x - 6)$ $(x - 3)(x - 5)$ $(x + 8)(x + 2)$ $(x + 7)(x + 2)$

$(x - 9)(x + 8)$ $(x - 6)(x + 1)$ $(x + 7)(x + 4)$ $(x - 6)(x - 4)$

$(x + 5)(x - 3)$ $(x + 7)(x - 3)$ $(x - 4)(x + 2)$ $(x - 5)(x + 2)$

$2(x - 6)(x + 4)$ $3(x + 3)(x - 3)$ $(x^2 + 9)$ $(x - 16)(x + 3)$ $(x + 6)(x - 4)$

$\sqrt{25}$	$\sqrt{9x^2}$	$\sqrt{144}$	$\sqrt{36}$
$\sqrt{81}$	$\sqrt{196x^4}$	$\sqrt{x^8}$	$\sqrt{x^6}$
$\sqrt{27}$	$\sqrt{20}$	$\sqrt{54}$	$\sqrt{x^{10}}$
$\sqrt{48}$	$\sqrt{4}$	$\sqrt{100}$	$\sqrt{121}$

© CD

± 6	± 12	$\pm 3x$	± 5
$\pm x^3$	$\pm x^4$	$\pm 14x^2$	± 9
$\pm x^5$	$\pm 3\sqrt{6}$	$\pm 2\sqrt{5}$	$\pm 3\sqrt{3}$
± 11	± 10	± 2	$\pm 4\sqrt{3}$